CHALLENGES OF BIOLOGICAL AGING

Edward J. Masoro received both his Ph.D. in physiology and A.B. from the University of California at Berkeley. He chaired the Department of Physiology at the University of Texas Health Science Center at San Antonio (UTHSCSA) from 1973 to 1991 and then became Founding Director of the Aging Research and Education Center at UTHSCSA. Currently a Professor Emeritus, he is director of the Research Development Core of the Nathan Shock Center for Excellence in the Biology of Aging at UTHSCSA. He was chair of the Department of Physiology and Biophysics at the Medical College of Pennsylvania from 1964 to 1973. Prior to that, he taught physiology at the University of Washington, Tufts University School of Medicine, and Queen's University (Canada).

He has served as President of the Gerontological Society of America, Chair of the Board of Scientific Counselors of the National Institute on Aging, editor of the *Journal of Gerontology: Biological Sciences,* and President of the Association of Chairmen of Departments of Physiology. His research in biological gerontology has focused on the antiaging action of caloric restriction in rodent models, for which he received the Allied-Signal Achievement Award in Aging, the Robert W. Kleemeier Award, the Irving Wright Award of Distinction, and the Glenn Foundation Award.

Challenges of Biological Aging

Edward J. Masoro, Ph.D.

Springer Publishing Company
New York

Copyright © 1999 by Springer Publishing Company, Inc.

All rights reserved

No part of this publication may be reproduced, stored in a retrieval system, or transmitted in any form or by any means, electronic, mechanical, photocopying, recording, or otherwise, without the prior permission of Springer Publishing Company, Inc.

Springer Publishing Company, Inc.
536 Broadway
New York, NY 10012-3955

Acquisitions Editor: Bill Tucker
Production Editor: Helen Song
Cover design by James Scotto-Lavino

99 00 01 02 03 / 5 4 3 2 1

Library of Congress Cataloging-in-Publication-Data

Masoro, Edward J.
    Challenges of biological aging / Edward J. Masoro.
        p.   cm.
    Includes bibliographical references and index.
    ISBN 0-8261-1277-3 (hbk.)
    1. Aging—Physiological aspects.   I. Title.
QP86.M27   1999
612.6'7—dc21   99-25949
                                                                                                                           CIP

Printed in the United States of America

Contents

Preface *xi*
Acknowledgments *xiii*

1 AGING: A BIOLOGICAL PUZZLE

THE PUZZLE	1
MORTALITY RISK	2
ANATOMICAL AND PHYSIOLOGICAL CHANGES	3
AGE-ASSOCIATED DISEASES	4
CONCEPT OF BIOLOGICAL AGE	5
GENETICS	6
UNIVERSALITY	10

2 DEMOGRAPHY AND THE SOCIETAL CHALLENGE

POPULATION MORTALITY DATA	14
Life Tables	14
Survival Curves	18
Age-Specific Mortality and Population Aging	18
FACTORS DETERMINING POPULATION AGE STRUCTURE	22
Changes in Life Expectancy	23
Changes in Birth Rate	24
Changes in Gender Composition	25
SOCIETAL IMPACT OF CHANGING AGE STRUCTURE	27

Work Force	27
Disease	28
Daily Living Assistance	28
NEED FOR BIOMEDICAL BREAKTHROUGHS	29

3 WHY AGING OCCURS

EVOLUTIONARY BIOLOGY	33
HISTORY OF EVOLUTIONARY THEORIES OF AGING	37
CURRENT EVOLUTIONARY THEORY OF AGING	38
GENETIC MECHANISMS IN THE EVOLUTION OF AGING	41
EVIDENCE SUPPORTING THE EVOLUTIONARY THEORY	42

4 HOW AGING OCCURS

OVERVIEW	47
THEORIES, HYPOTHESES, AND CONCEPTS	48
Evolutionarily Adaptive Aging Clocks	49
Wear and Tear	50
Rate of Living Theory	51
Free Radical Theory	52
Oxidative Stress Hypothesis	53
Glycation and Glycoxidation Hypotheses	55
Concept of Inadequate Protein Turnover	57
Concept of Altered Biological Membranes	57
Concept of Altered Extracellular Matrix	58
Genes and Gene Expression	59
Somatic Mutation Theory	60
Concept of Mitochondrial DNA Damage	62
Telomere Senescence Theory	63
Concepts Based on Altered Gene Expression	64
Regulation of Systemic Function	67
Endocrine and Neural Regulation	68
Altered Immune Function	71
CONCLUSIONS	72

Contents

5 BIOLOGICAL BASIS OF AGING: A UNIFYING CONCEPT

EVOLUTIONARY THEORY AS A GUIDE	76
THE DISPOSABLE SOMA THEORY OF AGING	77
"PRIVATE" AND "PUBLIC" PROXIMATE MECHANISMS OF AGING	77
UNIFYING CONCEPT	79

6 THE HUMAN AGING PHENOTYPE

BODY STRUCTURE AND COMPOSITION	86
Height	86
Weight	86
Lean Body Mass	86
Fat Mass	87
Cellularity	87
Cell Atrophy and Loss	88
Hypertrophy and Hyperplasia	88
Neoplasia	88
Extracellular Matrix	88
SKIN	89
Intrinsic Aging	90
Extrinsic Aging	91
Photoaging	92
Smoking	92
Age Changes in Skin Physiology	92
Skin Cancer	93
Pressure Sores	93
MUSCULOSKELETAL SYSTEM	94
Skeletal Muscle	94
Age Changes in Muscle Structure and Function	95
Age and Muscle Usage	95
Neuromuscular Function	95
Bone	96
Aging and Bone Loss	97
Osteoporosis	97
Risk Factors for Osteoporosis	98
Joints	98
Osteoarthritis	100

Rheumatoid Arthritis	100
Gout	101
NERVOUS SYSTEM	**101**
General Nervous System Age Changes	104
Structural Changes	105
Blood Flow and Metabolism	106
Sensory Functions	107
Skin Sense	107
Proprioception	108
Hearing	108
Vision	110
Taste and Smell	113
Motor Functions	113
Motor Responses to Sensory Input	114
Posture and Balance	114
Locomotion	115
Age-Associated Motor Disorders	116
Autonomic Nervous System	117
Sleep	118
Cognitive Functions	119
Attention	119
Memory	119
Intellectual Functions	121
Age-Associated Cognitive Disorders	121
CARDIOVASCULAR SYSTEM	**124**
Heart	125
Conductile System	125
Pump Function	126
Neural Regulation	127
Coronary Heart Disease	128
Congestive Heart Failure	129
Arteries	129
Arterial Structure	130
Blood Pressure	131
Microvasculature	132
Resistance Vessels	132
Exchange Vessels	133
Venous System	133
RESPIRATORY SYSTEM	**134**
Lung-Thorax Pump	134
Gas Transport Between Alveoli and the Tissues	136

RENAL AND URINARY SYSTEM — 137
- Renal Blood Flow — 138
- Glomerular Filtration — 140
- Tubular Functions — 140
- Regulatory Functions — 141
- Urination — 144
  - *Urinary Incontinence* — 145
  - *Prostatic Obstruction* — 146

GASTROINTESTINAL SYSTEM — 146
- Motor Activity — 146
  - *Mastication* — 148
  - *Swallowing* — 148
  - *Gastric and Intestinal Motility* — 149
  - *Colon Motility and Defecation* — 149
- Secretion — 150
- Digestion — 152
- Absorption — 152

ENDOCRINE AND METABOLIC FUNCTION — 153
- Energy Metabolism — 154
- Carbohydrate and Fat Metabolism — 155
- Protein Metabolism — 157
- Stress — 158
- DHEA (Dehydroepiandrosterone) and Melatonin — 159

FEMALE REPRODUCTIVE SYSTEM — 160
- Fertility — 161
- Menopause — 161
  - *Age of Occurrence* — 161
  - *Endocrine Changes* — 162
  - *Changes in Secondary Sex Organs* — 162
  - *Symptoms* — 162
  - *Hormone Replacement Therapy* — 163

MALE REPRODUCTIVE SYSTEM — 163
- Age Changes in the Testes — 164
- Age Changes in the Secondary Sex Organs — 164

IMMUNE SYSTEM — 165
- Age Changes in the Lymphoid Tissues — 167
- Age Changes in Lymphocytes — 167
- Disease and the Aging Immune System — 167

THERMOREGULATION — 168

7 POSSIBLE INTERVENTIONS TO RETARD AGING

DIETARY RESTRICTION IN RODENTS	171
PROPOSED PHARMACOLOGICAL INTERVENTIONS	175
Antioxidants	175
Deprenyl	177
Dehydroepiandrosterone (DHEA)	177
Melatonin	178
Growth Hormone	179
Estrogen	179
Testosterone	181
LIFESTYLE AND ENVIRONMENTAL FACTORS	181
Exercise	182
Diet	183
Smoking	185
Education and Social Support	186
Index	*189*

Preface

Gerontology, the science dedicated to the study of aging, encompasses many disciplines: biology, medicine, nursing, psychology, sociology, social work, economics, law, and political science, to name a few. However, because aging is fundamentally a biological process, an understanding of the biology of aging is needed to work most effectively in any of the fields that comprise gerontology.

The aim of this book is to provide the reader with an overview of what we know and, probably more importantly, what we do not know about the biology of aging. Why does aging occur? The emerging concepts of the evolution of aging have begun to be utilized in solving this puzzle, and these concepts are presented in this volume. What is the biological basis of aging? Many theories and hypotheses regarding the molecular, cellular, and systemic processes that may underlie biological aging are examined. This information is then synthesized in what I believe to be a unifying concept of the biological basis of aging.

Human aging is characterized by many anatomical and physiological changes as well as age-associated diseases. It is important to emphasize, however, that there is a marked variation among individuals in regard to the extent of a specific change. Many interventions, ranging from lifestyle modifications to pharmacological treatments, have been proposed to retard or even reverse age changes. The characteristics of these changes and interventions to modulate them are presented, and the latter are critically evaluated. Demographic changes in population structure projected for the 21st century pose enormous problems, and the potential role of research on the biology of aging in enabling society to meet these challenges is considered.

While this book is primarily intended for students training to become gerontologists, I believe that practicing gerontologists may also gain new insights in its pages. In addition, the book provides biologists who are not conversant with the biology of aging (i.e., most biologists) with a summary of the current state of knowledge of this subject. I am hoping that the biological puzzle of aging may lead some of them to investigate this under-studied area.

Although the text is not documented with citations from the original literature, a list of "Additional Reading" is provided at the end of each chapter, and documentation of most statements in the text can be found in the references cited in the listed books and articles. The *Additional Reading* lists will also enable the reader to pursue at greater depth subjects of particular interest.

It is my hope that this book will provide gerontologists in all disciplines with the biological insights needed to best accomplish their work. It is also my hope that it will convince biologists in general that aging is an intriguing aspect of biology and may well be a career avenue worth pursuing.

EDWARD J. MASORO

Acknowledgments

I am indebted to my wife, Barbara Weikel Masoro, for her help in editing this book.

1
Aging: A Biological Puzzle

When referring to living organisms, *aging* usually denotes not just the passage of time, but deterioration with the passage of time. *Senescence* is a more appropriate word for this usage, because aging can also refer solely to the passage of time. Senescence is defined as the *deteriorative changes, during the adult period of life, which underlie an increasing vulnerability to challenges, thereby decreasing the ability of the organism to survive*. Unfortunately, senescence is not a term widely used by the public, probably because of the feeling that it is synonymous with the adjective *senile* and the noun *senility*. In point of fact, senile and senility refer to the extreme deterioration of old age, the end result of the long-term progression of the processes of senescence. For instance, the loss of bone during the fourth decade of human life is an example of senescence, but a 40-year-old person can hardly be considered senile, even in regard to his or her bones. In this book, aging will be used in the sense of senescence, and the terms will be employed interchangeably.

To reiterate, the important point is that senescence refers not just to the passage of time, but to the occurrence of deteriorative processes with time. Thus, if no deterioration has occurred over a time interval, aging in the sense of senescence has not occurred.

THE PUZZLE

Based on what happens with the passage of time to inanimate objects, such as the appliances we use and the houses we live in, senescence in animals and plants does not seem at all surprising. Indeed, the Second Law of Thermodynamics states that with the passage of time, there is an increase in entropy; i.e., things become increasingly disordered. However, the Second Law of Thermodynamics applies to what physical chemists call a closed system, i.e., one in which the system in question can not exchange matter with its environment. Living organisms are not closed systems, but rather they are thermodynamically open systems; i.e., there is an exchange of matter between the living organism and its environment. Specifically, living organisms utilize the matter and energy in food (and light, in the case of photosynthetic organisms) to counter entropy. A striking example of this counterentropic use of external matter and energy is an animal's development from a fertilized egg to the highly complex young adult organism composed of a large number of cells with different kinds of specialized functions. The maintenance of this complex adult organism would appear to be a far less formidable challenge than its development. Yet even when the supply of external matter and energy is unlimited, it appears that most, if not all, species of complex organisms are not able to prevent deterioration of their structure and function during the adult period of life.

Aging is clearly not a thermodynamic inevitability, and yet it occurs widely in nature. Therefore, it must be concluded that the ability to utilize external matter and energy to maintain complex function and structure is progressively lost by most animals and plants with advancing adult calendar age. It is this conclusion that leads to the crux of the puzzle: *Why don't evolutionary forces eliminate a process, such as senescence, that is so detrimental to the ability of the individual organism to maintain itself and generate progeny?*

Considerable progress has been made towards solving this biological puzzle. However, before discussing this progress in depth in ensuing chapters, it is necessary to present a synopsis of what we do and do not know about aging.

MORTALITY RISK

With advancing adult calendar age in humans and many other species, the risk of dying increases. The extent of the increase in risk of mortality is assessed by determining the increase in *age-specific death rate* with increasing calendar age. The age-specific death rate (also called the age-specific mortality rate) refers to the fraction of the population of a given age that dies during a time interval (e.g., a 1-year period, such as 20–21 or 71–72 years of age). It is felt that the increase in the age-specific death rate with advancing calendar age is an index of increasing vulnerability due to senescence. Expressing mortality data in terms of age-specific death rate has the advantage of enabling meaningful comparisons to be made between populations of different sizes. For example, the number of people who die between the age of 50 and 51 is greater for the United States than for Canada simply because the United States has a much larger population; however, the age-specific death rate shows that the fraction of the Canadian and American population who die in each age interval is similar. The age-specific death rate is particularly useful when studying a *cohort* of animals in a laboratory setting. A cohort is a group of individuals that share a statistical characteristic. In the case of the animals in gerontologic laboratory studies, a cohort usually refers to individuals born during a particular period of time, such as a specific day in a particular calendar year. As the calendar age of a cohort increases, the number of individuals in that cohort decreases due to deaths. Thus because of the small number of animals reaching advanced ages, the number of individuals dying during an old-age interval may be less than during a young-age interval; however, assessment of the age-specific death rate usually shows that a much greater fraction of individuals entering an old-age interval dies compared to those entering a young-age interval, thereby revealing an increase in the vulnerability of the population with increasing age.

ANATOMICAL AND PHYSIOLOGICAL CHANGES

In addition to an increasing risk of mortality, many anatomical and physiological changes occur in humans and other species with increasing adult calendar age. Most of these changes appear to be deteriorative in nature. However, the extent to which such changes are the *result* or the *cause* of the age-associated increase in the vulnerability of organisms is not known and undoubtedly varies with the particular anatomical or physiological change. These age changes have been better studied in humans than in any other species, and an in-depth consideration of such information in regard to humans is the subject of Chapter 6. However, two general points should be considered here.

First, there is a great deal of individual variation in the occurrence and the magnitude of age changes in anatomical structures and physiological systems. J. W. Rowe, a geriatrician, and R. L. Kahn, a social psychologist and gerontologist, have pointed out that in a population of elderly humans, a subgroup can be found that shows minimal change with age in a particular physiological function; e.g., although most people exhibit a decline in kidney function with advancing age, a small but significant number of people do not show this decline. Rowe and Kahn have suggested that those elderly who show little deterioration in a constellation of physiological functions should be regarded as having undergone "successful" aging in physiological terms.

The second general point is that most elderly people do exhibit substantial deterioration in many physiological systems, and as a consequence, most of them have a reduced ability to cope with challenges and a reduced functional capacity. For example, most people with advancing calendar age have a reduced ability to successfully deal with the challenge of high or low environmental temperatures. Also, the capacity to carry out exercise is reduced with advancing age, a loss that ranges from an inability to walk in those with severely limited function, to an increase in the time it takes to run a marathon in the physically fit who have been marathon runners most of their lives.

AGE-ASSOCIATED DISEASES

There are many diseases that occur only at advanced ages or are more prevalent in the old. These are called age-associated diseases, and they underlie much of the increase in vulnerability seen with increasing adult calendar age. They are also responsible for much of the age-related change in the physiological systems. Age-associated diseases have been shown to occur in all animal species for which life span pathological analyses are available. For example, in rats, whose maximum length of life is about 4 years, cancers primarily occur after 2 years of age, while in humans, with a maximum length of life of about 100 years, they occur mostly after 50 years of age.

Moreover, each species has a set of age-associated diseases that differ from those found in other species. In humans, the increases in morbidity (a disease state) and mortality due to coronary heart disease, stroke, Type II diabetes, osteoporosis, Alzheimer's disease, Parkinson's disease, and prostatic cancer relate to calendar age in a fashion similar to that of the increase in the age-specific death rate. Although multiple sclerosis, amyotrophic lateral sclerosis, gout, peptic ulcer, and most cancers are also age-associated, their occurrence does not parallel the increase in the age-specific death rate.

CONCEPT OF BIOLOGICAL AGE

Since senescense involves not only the passage of time, but the deterioration of the organism with the passage of time, it is clear that calendar age does not provide an appropriate measure of senescence. To address this issue, gerontologists developed the concept of biological age as distinct from calendar age. To illustrate this concept, let us consider the case of individuals A and B who are of the same calender age (both 40-year-old men) but who differ markedly in the extent of senescent deterioration (individual A exhibiting much more deterioration than individual B). Although these two individuals are the same calendar age, individual A is considered much older in biological age than individual B.

While most gerontologists subscribe to this concept, the lack of methods for measuring biological age has undermined its usefulness for both the gerontologic researcher and the physician practicing geriatric medicine. In all fields of biology, the quantitative measurement of a process is usually key to a full understanding of the process. The ability to quantitatively measure the rate of aging of individual organisms would aid immeasurably in our quest to understand the basic nature of aging and in our efforts to develop interventions aimed at mitigating the negative consequences of aging. Intuitively, one would feel that developing such a measure would not be a great challenge but, as of now, an agreed-upon method is not available.

Indeed, much effort has gone into the development of biomarkers of aging as tools for the assessment of biological age. Biomarkers of aging are those changes with time (usually anatomical or physiological changes) in an organism that can serve as an index of the extent of senescent deterioration and can be readily measured in a relatively noninvasive manner. Specific examples of measurements that have been studied as potential biomarkers of aging include handgrip strength, speed of reaction to a stimulus, and the rate of growth of fingernails, to name a few. The measurement of vital capacity—the maximum volume of air that can be expired following a maximum inspiration—has shown the most promise as a potential biomarker of human aging. However, it has become clear that no single biomarker is likely to be indicative of the broad spectrum of changes occurring during senescent deterioration and, thus, it is unlikely that any single biomarker can provide a valid assessment of biological age.

Recognizing this problem, some investigators have responded by organizing a variety of potential biomarkers into what are called panels of biomarkers, and they have also developed mathematical models to estimate biological age from these panels. As of now, there is much disagreement about the validity of the panels and mathematical models because of the absence of reliable standards by which to evaluate them. One standard that has been suggested is the ability to predict the remaining length of life of the individual (i.e., the ability to predict individual mortality risk). Although this is a conceptually appealing suggestion, thus far, calendar age has proven to be a better predictor of remaining length of life than any of these models of biological age. It is not clear as to whether the suggested standard is inappropriate or the putative panels of biomarkers are faulty. Nevertheless, until appropriate standards are developed and agreed upon for testing the validity of putative biological markers of aging, use of these markers to measure the biological age of an individual will be open to question.

GENETICS

It is often said that if you want a long life, choose the right parents. Genetics does indeed play an important part in aging, but so do environment and gene-environment interactions. Much past and current research has been and is aimed at defining the role of genetics in aging, but before considering the results of this research, a brief overview of genetics is needed.

The genetic information in the cells of an organism is called the genome. Genes are the functional units of the genome. Chemically, genes are components of DNA molecules; and in plants and animals, the DNA molecules reside primarily in the chromosomes of the nuclei of cells. Each chromosome has a single DNA molecule which contains many genes (about 3,000 per human chromosome).

The term phenotype refers to the observable characteristics of an organism, such as size, running ability, musical talent, etc.; it is determined both by the genes and by the influence of a variety of environmental factors, such as nutrition, physical training, and education. Although humans and rats have many of the same genes, the divergence of the two species over the mega-years has been the result of the accumulation in each species of many small genetic modifications, resulting in genes unique to each species. It is this relatively small difference in their genes that plays the major role in the marked difference in phenotype between humans and rats. However, in the case of the smaller phenotypic differences among individuals within a species, the environment and its interaction with the genes also have an important role. For example, there has been a significant increase over the past 50 years in the height of men in Japan, which cannot be due to genetics *per se,* but rather must result from environmental changes and their interaction with the genes. Prior to World War II, it was known that children of the Japanese who migrated to the United States were taller than the members of their generation born and reared in Japan. Thus it seems likely that the considerable postwar American influence on Japanese life played a large role in the increase in height of Japanese men. Of the many aspects of life so influenced, the postwar change in the diet of the Japanese (i.e., the "Westernization" of the diet) seems likely to be the factor most responsible for the increase in height.

Aging is a phenotypic characteristic that varies greatly among species in its rate of progression; certainly, the fact that the maximum life span for rats is about 4 years, compared to about 100 years for humans, is evidence of this fact. Genetics, not environment, is undoubtedly the major factor for the great differences in the rate of aging among species.

The reasons for the differences in rate of aging among individuals within a species are complex. The structure of the DNA of a particular gene can vary slightly among individuals of the same species, and these different versions of the same gene are called *alleles*. Some alleles will always result in a particular phenotypic characteristic, as is the case with the allele determining the color of brown eyes. The phenotypic effects of other alleles depend on the environment and the rest of the genetic makeup of the individual. Most cells have two copies of each chromosome, which are called homologous chromosomes because each has the same genes; one of the homologous chromosomes is derived from the mother and the other from the father of the individual. If a particular allele of a gene is present in both homologous chromosomes, the individual is homozygous for that allele; and if the allele is present in only one of the homologous chromosomes, the individual is heterozygous for the allele.

Is there evidence that differences in alleles among individuals of a species influence the rate of aging? To address this question, the heritability of this trait must be determined; "heritability" refers to the fraction of the variability in a particular characteristic that results from genetic effects, rather than from environment or chance. Estimates of the fraction of the variation in human length of life due to genetics have been made, based on analyses of parent-offspring correlations in length of life and differential lengths of life of identical and fraternal twins. These estimates of heritability of longevity have varied greatly, from as low as 10% to as high as 70%. Moreover, the validity of using individual longevity as an index of the rate of aging is also open to debate. Thus, at this point in time, the heritability of the rate of aging in humans is not precisely known.

Another approach to learning about the influence of genetics on the rate of aging is to study individuals with certain genetic diseases. Some, but not all, aspects of the aging phenotype are accelerated in several human genetic syndromes (e.g., Hutchinson-Gilford Syndrome, Werner Syndrome, and Down Syndrome). George M. Martin has introduced a term, segmental progeroid syndromes, to refer to these syndromes because not all aspects of the aging phenotype are accelerated. Hutchinson-Gilford Syndrome, in which aspects of the senescent phenotype appear in children, is extremely rare (about 20 cases have been recognized worldwide); these children show balding, the facial features of an old person, and atherosclerosis, and they usually die before 15 years of age, most often from a heart attack. (Interestingly, they have normal intelligence.) Werner Syndrome, which makes its appearance during adolescence, is also rare, but less rare than Hutchinson-Gilford Syndrome. Those suffering Werner Syndrome have a premature occurrence and rapid progression of atherosclerosis and also have skin changes that give the appearance of an old person. They are of normal intelligence and usually die in their 30s. The genetic defect in Werner Syndrome has been identified; it is a *mutant* (altered) form of a gene of the helicase family (an enzyme family involved in the functioning, repair, and replication of DNA). Down Syndrome is relatively common compared to the other two syndromes and is characterized by the premature appearance of many aspects of the aging phenotype, such as an early decline in immune function and the early appearance of cancers, cataracts, and degenerative vascular diseases. Almost all such patients show the pathologic signs of Alzheimer's disease if they live into their 40s. Down Syndrome appears to be the result of an extra copy of chromosome 21. Although these studies of segmental progeroid syndromes have provided very interesting findings, the relevance of this information to aging as it occurs in most people remains an open question.

Still another approach to the study of the effects of gene alterations on the rate of aging is to experimentally cause single gene mutations (alterations in the structure of a single gene) in species in which the genome can be readily manipulated. In species of yeast and nematodes, single gene mutations have been produced that increase the length of life, presumably by slowing the rate of aging.

Based on these many different approaches, it can be concluded that genetics is the major factor in species differences in the rate of aging. Moreover, laboratory studies show that genetics can play a major role in the individual differences in the rate of aging within a species. However, environment and gene-environment interactions are also important factors in the differences in the rate of aging among the individuals of a species. The laboratory studies on the effects of restricting the food intake of rodents provide clear evidence of the importance of diet. Reducing the food intake of rodents markedly increases the length of life and postpones most aspects of the aging phenotype. Honeybees provide a remarkable example of the influence of lifestyle on life span and thus presumably on the rate of aging. Female worker bees and the queen bee have very different lifestyles, and even when genetically identical, they have markedly different life spans (2 to 8 months for the workers, about 5 years for the queen). Thus, marked differences in the rate of aging among individuals can result from diet, lifestyle, and other environmental factors rather than genetics.

UNIVERSALITY

Do all species undergo senescence? This is a difficult question to answer. The first issue is that of criteria for determining the occurrence of senescence. An increase in the age-specific death rate with increasing adult calendar age is one criterion that has been used, and the other is decreasing fecundity with increasing calendar age. However, so few species have been assessed for these criteria that an answer to the question of universality cannot be given with anything near certainty. Indeed, rock fish, sturgeons, and carp appear to live to advanced ages without evidence of an increase in age-specific death rates or a decrease in fecundity. However, it is quite possible that these species have not been studied carefully enough to detect such changes if, in these species, they progress at an extraordinarily slow rate.

In summary, empirical studies addressing the question of which species undergo senescence have been done with only a few species. Nevertheless, biologists have long believed that senescence occurs in those species with a germ line (the lineage of cells that produce the gametes, i.e., sperm and ova) that is separate from the soma (all cells other than those of the germ line) and that it is only the soma that undergoes senescence. However, recent findings have shown that senescence can occur in organisms that do not have a germ line separate from the soma, such as some species of yeast. Indeed, our current understanding of the evolutionary biology of aging (see Chapter 3) leads to the conclusion that senescence probably occurs in most, if not all, species that reproduce sexually.

ADDITIONAL READING

Brown, W. T. (1991). Genetic diseases of premature aging. V. J. Cristofalo, (Ed.). *Annual Review of Gerontology and Geriatrics, 10,* 23–42.

Ferraro, K. F. (Ed.). (1997). *Gerontology: Perspectives and issues* (2nd ed.). New York: Springer.

Finch, C. E., & Tanzi, R. E. (1997). Genetics of aging. *Science, 278,* 407–411.

Martin, G. M. (1996). Genetic modulation of the senescent phenotype of Homosapiens. *Experimental Gerontology, 31,* 49–59.

Martin, G. M. (1997). Genetics and the pathobiology of aging. *Philosophical Transactions of the Royal Society, London,* B, 352, 1773–1780.

McClearn, G. E. (1997). Biomarkers of age and aging. *Experimental Gerontology, 32*, 87–94.
Ricklefs, R. E., & Finch, C. E. (1995). *Aging: A natural history.* New York: Scientific American Library.
Rowe, J. W., & Kahn, R. L. (1987). Human aging: Usual and successful. *Science, 237*, 143–149.
Schacter, F. (1988). Causes, effects, and constraints in genetics of human longevity. *American Journal of Human Genetics, 62*, 1008–1014.

2
Demography and the Societal Challenge

Aging has long been a subject of great human interest, as is evident from Shakespeare's detailed description of the progression of the aging phenotype in *As You Like It:*

> The sixth age shifts into lean and slipper'd pantaloon, with spectacles on nose and pouch on side, his youthful hose well sav'd, a world too wide, for his shrunk shank; and his big manly voice, turning again toward childish treble, pipes and whistles in his sound. Last scene of all, that ends this strange eventful history, is second childishness, and mere oblivion, sans teeth, sans eyes, sans taste, sans everything.

Only recently, however, has aging become a major political and economic issue. This recent public interest stems from the demographic change that has occurred during the last half of the 20th century. Prior to that time, the elderly accounted for a small fraction of the human population; thus, the societal impact of aging was small. Since the middle of the 20th century, however, the elderly have become a progressively increasing fraction of the world population, a trend projected to continue well into the 21st century.

It is the age structure of the population that has economic and social consequences. In 1900, only about 4% of the population in the United States was 65 years of age or older; by 1978, it was 11%, further increasing to 13% by 1990.

A similar increase has occurred in most developed nations. In 1950 in such nations, the fraction of the population over 60 years of age was 11.4%, and over 80 years of age, 1%; by 1990, this had increased to 17.1% and 2.6% respectively.

In the developing nations, in 1990, only 6.9% of the population was over 60 years of age, and 0.5% over 80 years of age; while still only a small fraction of the population, it is a significant increase compared to 1950.

It is projected that the fraction of the elderly in the world population will continue to increase well into 21st century, with an estimate that in 2025 more than 25% of the population of the developed nations will be over the age of 60, with about 12% in the case of the developing nations. A graphic view of the age structure of the population of the United States in 1955 and a projection of what it will be in 2010 are shown in Figure 2.1.

POPULATION MORTALITY DATA

Before assessing the factors underlying past as well as projected future changes in population age structure, it is important to consider population mortality data. Such data not only play a key role in the work of demographers but are also of great use to those working in many research areas of gerontology.

Life Tables

Data relating calendar age and population mortality are collected in what are called life tables. Table 2.1, in which the data for a cohort of male houseflies are reported, is an example of such tables. The first vertical column (designated x) contains the age intervals chosen by the tabulator. In this case, where the longest life is less than 60 days, intervals of 1 day duration were chosen with the first interval (0–1) referring to the first day of life. (In the case of human populations with a maximum life span of more than 100 years, a 1-year age interval is usually chosen.) The second vertical column (designated d_x) records the number of houseflies that died during each age interval. (The reader may find it confusing that a fraction of a fly is reported to have died as in the 1.5 in the first entry in Table 2.1; the explanation is that for convenience the data reported are per 1,000 flies but the initial size of the population was 4,627 flies; thus 7 flies died, or 1.5 per 1,000 flies.) The third vertical column (designated l_x) lists the number of flies alive at the beginning of the age interval (again, for convenience, the number in Table 2.1 is per 1,000 flies). The fourth

Demography and the Societal Challenge

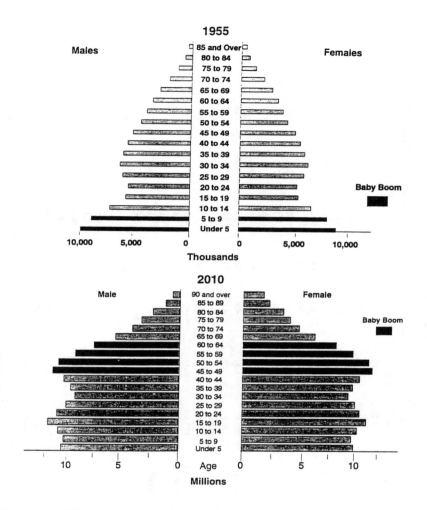

Figure 2.1 The age composition of the United States population for the year 1955 and a projection for the year 2010.

From *Sixty-Five Plus in America* (Current Population Reports: Special Studies, P23-178RV), pp. 2–4, 2–5, by C. M. Taeuber, 1992, Washington, DC: U. S. Government Printing Office.

vertical column (designated q_x) contains the age-specific death rates, i.e., the fraction of houseflies entering an age interval that die during that interval (note that in Table 2.1 the age-specific death rate is multiplied by 1,000 in order to avoid tabulating small decimal fractions). The fifth vertical column (designated e_x) records the *life expectancy*.

Table 2.1 Life Table for 4,627 Male Houseflies *Musca domestica* L. (NAIDM Strain)

x	d_x	l_x	$1{,}000\,q_x$	e_x	x	d_x	l_x	$1{,}000\,q_x$	e_x
0–1	1.5	1000.0	1.5	16.88	30–31	8.4	49.1	171.1	5.11
1–2	2.8	998.5	2.8	15.90	31–32	5.6	40.6	137.9	5.07
2–3	3.7	995.7	3.7	14.94	32–33	6.7	35.0	191.4	4.80
3–4	2.8	992.0	2.8	14.00	33–34	4.3	28.3	151.9	4.82
4–5	2.6	989.2	2.6	13.04	34–35	5.0	24.0	208.3	4.59
5–6	5.6	986.6	5.7	12.07	35–36	3.0	19.0	157.9	4.66
6–7	8.6	981.0	8.8	11.14	36–37	2.6	16.0	162.5	4.44
7–8	14.0	972.3	14.4	10.23	37–38	1.5	13.4	111.9	4.21
8–9	19.5	958.3	20.3	9.37	38–39	2.4	11.9	201.7	3.67
9–10	48.0	938.8	51.1	8.56	39–40	2.6	9.5	273.7	3.47
10–11	46.5	890.9	52.2	7.99	40–41	2.8	6.9	405.8	3.59
11–12	57.5	844.4	68.1	7.40	41–42	1.1	4.1	268.3	4.71
12–13	94.9	786.9	120.6	6.91	42–43	0.6	3.0	200.0	5.23
13–14	69.8	692.0	100.9	6.79	43–44	0.2	2.4	83.3	5.42
14–15	77.8	622.2	125.0	6.49	44–45	0.2	2.2	90.9	4.86
15–16	64.2	544.4	117.9	6.35	45–46	0.4	1.9	210.5	4.53
16–17	62.0	480.2	129.1	6.13	46–47	0.4	1.5	266.7	4.60
17–18	61.8	418.2	147.8	5.96	47–48	0.0	1.1	0.0	5.09
18–19	52.7	356.4	147.9	5.91	48–49	0.0	1.1	0.0	4.09
19–20	49.9	303.7	164.3	5.85	49–50	0.4	1.1	363.6	3.09
20–21	41.7	253.7	164.4	5.90	50–51	0.2	0.6	333.3	4.17
21–22	31.8	212.0	150.0	5.97	51–52	0.0	0.4	0.0	5.00
22–23	28.1	180.2	155.9	5.93	52–53	0.0	0.4	0.0	4.00
23–24	22.9	152.2	150.5	5.93	53–54	0.2	0.4	500.0	3.00
24–25	16.6	129.2	128.5	5.90	54–55	0.0	0.2	0.0	4.50
25–26	14.0	112.6	124.3	5.69	55–56	0.0	0.2	0.0	3.50
26–27	14.9	98.6	151.1	5.43	56–57	0.0	0.2	0.0	2.50
27–28	18.8	83.6	165.1	5.32	57–58	0.0	0.2	0.0	1.50
28–29	12.3	69.8	176.2	5.27	58–59	0.2	0.2	1000.0	0.50
29–30	8.4	57.5	146.1	5.29					

Note. From "A life table for the common housefly, *Musca domestica*," by H. Rockstein and H. M. Lieberman, 1959, *Gerontologia*, 3, pp. 23–26. Copyright 1959 by S. Karger AG, Basel, Switzerland. Reprinted with permission.

Life expectancy is usually used in newspapers and magazines to refer to the mean length of life projected for a population of newly born individuals (e.g., Americans). However, it is more broadly defined as the mean remaining length of life projected for a population of humans or any other species entering any given age interval. It is this broader definition that is used by insurance actuaries. Life expectancy when referring to humans is an estimated projection unless a human cohort born well over 100 years ago is being considered. On the other hand, in laboratory studies in which a particular cohort of animals with relatively short life spans (such as mice, rats, fruit flies, and nematodes) is examined until the last member dies (as is the case for the study reported in Table 2.1), life expectancy can be determined rather than estimated by projection.

Life tables often also contain data on survival rate and fecundity, information of great importance for those interested in the demography of aging. However, for now, our focus will be on mortality; the other issues will be discussed later in this chapter.

There are two kinds of life tables: cohort life tables and period life tables. A cohort life table contains data collected over the entire time span of a particular cohort. Table 2.1 is an example of a cohort life table. Such data permit life expectancies to be calculated rather than projectively estimated. Period life tables are used when it is not possible to readily follow a cohort until the last member dies, which is often the case for species with long life spans, such as humans. Indeed, human cohort tables are available, but they are incomplete except for cohorts born well over 100 years ago. In period life tables, the data for each age interval are from a different cohort at a given point in time (e.g., the cohorts making up the United States population in the year 1997). Because a period life table contains data from many different cohorts, it does not provide accurate information for any particular cohort. For example, in populations where the age-specific death rate has been progressively declining for an extended period of time, as has occurred in the United States during the 20th century, the projected life expectancies from data in a period life table are likely to be an underestimate for many of the cohorts comprising the table. The point to be emphasized is that life expectancies from period life tables are projections based on the mortality characteristics at the time of compiling the table, and thus may not be an accurate forecast of the actual remaining mean length of life of any of the cohorts in the table.

Survival Curves

Since the large data sets reported in life tables are cumbersome, they are not often found in gerontologic publications. Rather, the information contained in life tables is usually reported graphically. Some of the data are presented in what are called survival curves (*x axis*, age; *y axis*, percent of population alive), which are quite useful because of the readily comprehensible visual information they provide. The survival curves in Figure 2.2 were produced from cohort life tables generated from a study carried out in my laboratory with a cohort of male rats. The cohort was divided into two experimental groups: One, designated Group A, was fed *ad libitum* (i.e., allowed to eat as much as they wanted); the other, designated Group R, had their food intake restricted to 60% of the mean intake of Group A. The y axis in Figure 2.2 denotes the percentage of the starting rat population that is alive, and the x axis denotes calendar age in days of the population. These curves make it apparent at a glance that the population in Group R had a longer mean, median, and maximum length of life than rats in Group A.

The survival curves in Figure 2.3 were produced from period life tables for the United States population in 1910 and 1970. Clearly, a greater percentage of the population is projected to reach advanced ages in 1970 than in 1910. The change in shape of the survival curve is termed "rectangularization" because the 1970 curve looks more like two sides of a rectangle than does the 1910 curve. Indeed, survival curves of human populations have become progressively more rectangular throughout the 20th century, which is another way of stating that an ever-increasing fraction of the population is living to advanced ages.

Age-Specific Mortality and Population Aging

As discussed in Chapter 1, the age-specific death rate is viewed as a measure of the vulnerability of those in a particular age interval of a population. Based on the reasonable assumption that senescence causes vulnerability to increase, the concept emerged that the rate of increase of the age-specific death rate with increasing calendar age can serve as a quantitative index of the rate of population aging.

The influence of calendar age on age-specific death rate is most easily comprehended by presenting these life table data in a graphic

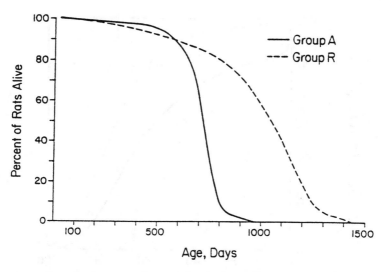

Figure 2.2 Survival curves from cohort life tables for population groups of male F344 rats. Group A rats were *ad libitum* fed and Group R rats were restricted to 60% of the mean food intake of Group A.

From "Life Span Study of SPF Fischer 344 Male Rats Fed *ad libitum* or Restricted Diets: Longevity, Growth, Lean Body Mass, and Disease," by B. P. Yu, E. J. Masoro, I. Murata, H. A. Bertrand, and F. T. Lynd, 1982, *Journal of Gerontology, 37*, 135.

format. This is done in Figure 2.4 for the United States populations of 1910 and 1970. In this graph, the x and y axes denote the age intervals and the age-specific death rate (log scale) respectively. The graphs in Figure 2.4 show a high age-specific death rate at birth, which falls until about 10 years of age, and then increases markedly during adolescence. The increased death rate during adolescence probably does not reflect increased organismic vulnerability but rather an increase in risk-taking behavior, leading to an increase in death due to accidents and social behavior. Of interest to gerontologists is the approximately exponential increase in age-specific death rate with increasing calendar age that starts at about 30 years of age (appearing in the semilog graphs in Figure 2.4 as a linear change in age-specific death rate with increasing calendar age). This exponential increase in age-specific death rate was first described by Benjamin Gompertz, a British actuary, in 1825. However, as the database on human age-specific death rates has enlarged, it has become clear that the increase in age-specific death rate with increasing

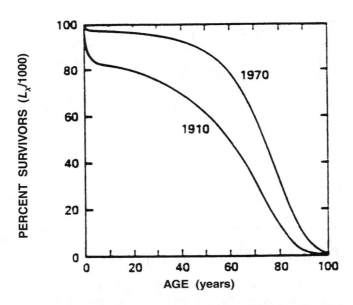

Figure 2.3 Survival curves from period life tables for the United States population (men and women) for the years 1910 and 1970.

From *Vitality and Aging* (p. 29), by J. F. Fries and L. F. Crapo, 1981, New York: W. H. Freeman and Company. Copyright 1981 by W. H. Freeman and Company. Reprinted with permission.

calendar age becomes less than exponential at very advanced ages, a phenomenon that greatly interests gerontologists and therefore will be discussed in greater detail later in this chapter.

Further examination of the data in Figure 2.4 provides an interesting comparison between the 1910 and 1970 populations of the United States. The age-specific death rate for individuals in the same age interval (e.g., 10–11, 35–36, 59–60 years of age) was much higher in 1910 than in 1970. However, the graphs for both 1910 and 1970 show a nadir at about 10 years of age and an exponential increase in age-specific death rate with increasing adult calendar age. What is most striking is that the rate of this exponential increase (i.e., the percent increase in age-specific death rate with increasing calendar age) is similar for both populations, even though the absolute age-specific death rate is higher in 1910. This finding suggests that the increase in vulnerability with adult age was similar in 1910 and 1970, but that individuals living in 1970 had a more protected environment than those living in 1910.

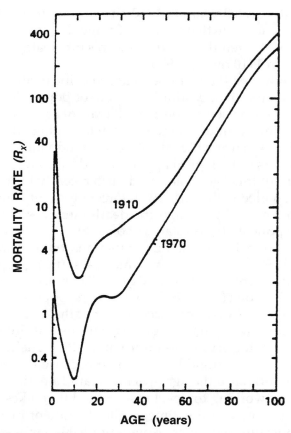

Figure 2.4 Age-specific mortality for the United States population (men and women) for the years 1910 and 1970.

From *Vitality and Aging* by J. F. Fries and L. F. Crapo, 1981, New York: W. H. Freeman and Company. Copyright 1981 by W. H. Freeman and Company. Reprinted with permission.

The amount of elapsed time before the age-specific death rate doubles, called the Mortality Rate Doubling Time (MRDT), inversely relates to the rate of increase in age-specific death rate. The MRDT is often used as an inverse index of the rate of aging. For example, the MRDT for humans, horses, dogs, and rats is 8, 4, 3 and 0.3 years respectively; based on this index, the hardly surprising conclusion is that the rate of aging of humans is slow and that of rats fast, with horses and dogs falling in between. The MRDT for humans in the

United States was 8 years in 1910 and 8 years in 1970, which has been taken to indicate that the rate of aging did not change between 1910 and 1970, even though the age-specific death rate at all ages was higher in 1910 than in 1970.

For many years, the increase in age-specific death rate has been the "gold standard" measure of the rate of population aging. This measure has been used to compare the rate of aging among species, to evaluate interventions aimed at slowing aging, and to evaluate methods for measuring the rate of aging of individuals, such as proposed markers of biological age discussed in Chapter 1. However, the validity of this "gold standard" has recently been questioned. Research reported in the early 1990s forcefully pointed to the fact that at advanced ages the age-specific death rate does not continue to increase exponentially. As stated above, this fact had long been recognized as occurring in human populations, but research with fruit flies in the 1990s found that in some cases, the age-specific death rate actually decreased at very advanced ages. Conventional thinking would lead one to conclude that in some populations the rate of aging decreases at very advanced ages, but this is not likely. Rather, it seems more likely that in some situations the "gold standard" method is yielding erroneous information. Piantanelli and his associates believe the method breaks down because factors other than senescence influence the age-specific death rate; these researchers are currently working to develop a method that takes into account such factors. Thus, at this time, while the rate of increase in age-specific mortality with increasing adult calendar age remains the best means available to measure population aging, the potential pitfalls of this method must always be kept in mind.

FACTORS DETERMINING POPULATION AGE STRUCTURE

The two factors that determine the age structure of the world population are life expectancy and birth rate. In the case of nations, there is the additional factor of migration into and out of the nation. Migration usually plays only a minor role, but it must be noted that there has been a recent high rate of migration of young people from Latin America and Asia, which could alter the current and future age structure of the United States. Since this factor and its potential

Changes in Life Expectancy

Given the progressive decrease at all ages in the age-specific death rate which has occurred during the 20th century, it is not surprising that life expectancy would have increased during this period. Indeed, there was a remarkable increase in life expectancy in the developed nations. In 1890, the life expectancy from birth in Europe and the United States was less than 47 years. By 1965, in the United States it increased to 67 years for men and 73 years for women; and by 1990, it was 73 years for men and 78 years for women. Similar increases in life expectancy have occurred in the nations of Western Europe. It is interesting to note that Japan has experienced the greatest increase in life expectancy: for men, life expectancy from birth rose from 68 years in 1965 to 76 years in 1990; and for women over that same time span, it rose from 73 years to 83 years. Moreover, significant increases in life expectancy from birth have also occurred for both men and women in the developing nations, rising from about 40 years in the early 1950s to almost 62 years in 1990. Further increases are predicted by the United Nations for the first part of the 21st century, with the life expectancy from birth in 2020–2025 for men projected to be about 76 years in the developed nations and almost 70 years in the developing nations, and for women about 82 years in the developed nations and almost 74 years in the developing nations.

Intuitively, one would feel that the increase in life expectancy (i.e., people on average living longer) during the 20th century has been a major reason for the increasing fraction of old people (i.e., people over 65 years of age) in the population. However, analyses of the data reveal that increases in life expectancy have not been a major factor. In point of fact, increases in life expectancy during the 20th century have resulted primarily from prevention of death from infectious diseases at young ages, thereby affecting all age strata of the population to almost the same extent. However, if life expectancy continues to rise in the 21st century, it will primarily influence the older age strata of the population, because currently, in developed countries, few deaths occur at young ages; increasingly, this is also the case in the developing nations. Indeed, in the developed nations, future increases in life expectancy will primarily be due to the prevention or delay of such age-associated diseases as cancer, coronary heart disease, and stroke. Since future increases in life expectancy will affect primarily the older strata of the population, such increases will play an important role during the 21st century in increasing the fraction of the population that is old, particularly the fraction classified as old-old (i.e., over 85 years of age).

Changes in Birth Rate

In the developed nations, industrialization and urbanization during the 19th century made it desirable by the end of that century for families to reduce the number of progeny. Indeed, there has been a continuing, progressive decrease in the birth rates (i.e., number of births per woman) in these countries during the 20th century, although to be sure there has been the occasional deviation from this trend, such as the high birth rate in the United States in the years immediately after World War II, generating what is referred to as the "Baby Boom Generation." In fact, the birth rate has decreased markedly in both developed nations and developing nations over the last half of the 20th century. During the period of 1950–1955, the birth rate in the developed countries was 2.8 per woman and in the developing nations it was 6.2 per woman; by the period of 1985–1990, the rate had fallen to 1.9 per woman in the developed nations and 3.9 per woman in the developing nations. The United Nations has projected a further decline in birth rate in the developing nations during the first part of the 21st century, with a birth rate of 2.3 per woman by 2025. There are many reasons for the reduction in the birth rate over the past 50 years in the developed nations; these include socioeconomic factors that promote family planning (e.g., the enormous increase of women in the work force), the increased access to effective contraceptives, and the legalization of abortions in some nations (e.g., in the Soviet Union, it played a major role). In the developing nations, the same factors, as well as official government policies favoring small family size, have been responsible for the reduction in birth rate that has occurred and that is projected to continue in the first part of the next century.

Thus it is the decrease in birth rate, not the increase in life expectancy, that is the reason for the marked rise in the 20th century in the fraction of the population that is elderly. Indeed, the great effect of birth rate on the age structure of a population is well illustrated by the "Baby Boom Generation" of the United States, as shown in Figure 2.1. The high birth rate, which yielded the "Baby Boom Generation," heavily weighted the population structure in 1955 toward young people. However, it is projected that in 2010 that same group will cause the population of the United States to be weighted toward middle-aged people, and by 2025 (not shown in the figure) that generation will result in a population weighted in favor of old people. In addition, as stated earlier, the projected further increase in life expectancy during the 21st century will be another factor in the production of populations with high percentages of elderly, particularly the old-old.

Changes in Gender Composition

In the 19th century, the life expectancy of men and women in the United States was similar in spite of the high incidence in women of death associated with pregnancy and, in particular, with childbirth. During the 20th century, life expectancy for women in the United States and other developed nations has increased to a greater extent than life expectancy for men. By 1950–1955, the life expectancy of women in the developed nations was 5.3 years longer than men, and in the developing nations it was 1.8 years longer. By 1985–1990, the difference had increased further to 7.1 years in the developed nations and 2.7 years in the developing nations. The United Nations projection for 2020–2025 is for the difference to increase further in the developing nations to 4.2 years, but to decrease somewhat in the developed nations to 6.1 years; based on recent trends, the projected further increase in life expectancy for men in the developed nations during these years is a little greater than for women, thus narrowing the gender difference.

The age-specific death rate (i.e., the fraction of the population entering an age interval that dies during that age interval) is graphically shown in Figure 2.5 for the 1976 population of Caucasian men and women in the United States. The striking finding that emerged from this analysis is that at all age intervals women had a lower age-specific death rate than men, but the rate of increase in the age-specific death rate with increasing calendar age and the mortality rate doubling time did not differ between men and women. These

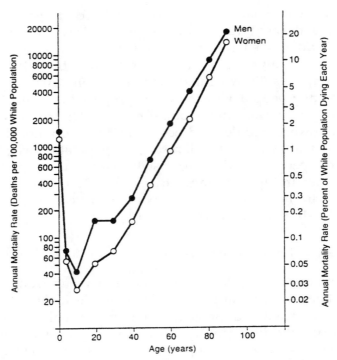

Figure 2.5 The age-specific mortality rate for Caucasian men and women of the United States in 1976.

From "Biological Basis of the Sex Differential in Longevity," by W. R. Hazzard, 1986, *Journal of the American Society of Geriatrics, 34,* p. 459. Copyright 1986 by Williams & Wilkins. Reprinted with permission.

findings suggest that men and women age at the same rate, but that men at all ages are more vulnerable than women of the same age.

Not surprising to anyone familiar with retirement communities, the gender difference in life expectancy results in many more women than men in the population of people of advanced age. In 1950, the population of those over 80 years of age was comprised of 63% women in the developed nations and 59.1% in the developing nations. By 1990, it increased to 70.5% women in the developed nations, while remaining almost unchanged at 59.2% women in the developing nations, because since 1950 the increase in life expectancy for men has been almost as great as for women in the developing nations.

SOCIETAL IMPACT OF CHANGING AGE STRUCTURE

There is great concern that the marked increase in the fraction of the population that is old, particularly the old-old, during the first part of the 21st century will have a negative impact on society. It is felt that there will be serious socioeconomic problems with political consequences and that both the developed and the developing nations will have to cope with these issues. This concern is based on the biological realities of aging and, in addition, on widely held perceptions of what the biological realities of aging may be.

Work Force

The age structure of the population is the major factor determining the fraction of the population that makes up the work force, i.e., the fraction of the population producing goods and services for the entire population. In the first decades of life, the individual does not participate in the work force, or does so to a limited extent. Then, after many decades in the work force, the individual retires. Economists view the two periods of life in which an individual is not in the work force as periods of dependency, in that the individual is not producing but is consuming goods and services. These periods are referred to as young-age dependency and old-age dependency. Dependency ratios are calculated to quantitatively assess the approximate extent of dependency in the population. The number in the population under age 21 divided by the number between the age of 20 and 65 provides the young-age dependency ratio. The number of people over 65 divided by the number between 20 and 65 provides the old-age dependency ratio. The total dependency ratio is, of course, the sum of the young-age dependency ratio and the old-age dependency ratio. In the United States in 1950, the young-age dependency ratio was 0.586, the old-age dependency ratio was 0.138, and the total dependency ratio was 0.724; in 1995, the young-age dependency ratio was 0.494, the old-age dependency ratio was 0.214, and the total dependency ratio was 0.708. It is projected that in the United States in 2025, the young-age dependency ratio will be 0.423, the old-age dependency ratio will be 0.318, and the total dependency ratio will be 0.741.

Economists tend to ignore the fact that many retired people provide goods and services as volunteers and unpaid helpers. In addition, many aged 65-plus have not retired, or have rejoined the work force on a part-time or full-time basis. Nevertheless, the amount of goods and services needed for the old is greater than for the young. The need for services is particularly great in the elderly. It is estimated that the cost of goods and services needed for an old person is about three times that for a young person. Thus, the increases in the old-age dependency ratio that have occurred in the last half of the 20th century and the further increases projected for the first part of the 21st century are, indeed, reasons for concern.

Disease

A major reason for the high costs involved in meeting the needs of old people is disease. As discussed in Chapter 1, many diseases are far more prevalent at advanced ages, and these diseases tend to be chronic, often progressing with increasing severity over a period of years. Some tend to be fatal, such as many cancers, ischemic heart disease, and cerebrovascular disease, while others do not result in death but interfere with the functioning of the individual over a long period of time (e.g., the various forms of arthritis, visual and hearing impairments, and gastrointestinal problems, to name a few). In the age range of 65 to 74 years, men have a higher prevalence than women of most fatal chronic diseases, while women have a greater prevalence than men of most nonfatal chronic diseases. In addition, the old are less able to cope with acute diseases, such as influenza, and thus require more medical support than the young afflicted with such diseases. Indeed, in 1993 in the United States, the admission to hospitals per 1,000 people was twice as great for those 65–74 years old and two and a half times greater for those 75 years of age and older than it was for those 45–64 years old; and the average length of stay in the hospital was 18% longer for those 65–74 years old and 31% longer for those 75 years old and above than for those 45–64 years old. Furthermore, with advanced age, the need for nursing home care and home health care programs increases even more markedly than hospital care.

Daily Living Assistance

With increasing calendar age past 65 years, an increasing percentage of the elderly lose the ability to carry out one or more of what are called Instrumental Activities of Daily Living (IADL) such as shopping for and preparing meals, managing money, doing laundry, using the telephone, doing heavy housework, doing light housework, and outside mobility. Also increasingly lost with advancing old age is the ability to do one or more of the most basic aspects of daily living, referred to as the Activities of Daily Living (ADL), such as using the toilet, bathing, getting in and out of bed, dressing, eating, and inside mobility. In the age range of 65 to 74 years, almost 12% have difficulty with one or more of the ADL, and of these, about three-quarters receive help in dealing with the deficit. For those 85 years of age and older, more than half have difficulty with one or more of the ADL, and of these more than 85% receive help in dealing with the deficit. Much of the IADL and ADL loss is due to chronic diseases, but other aspects of aging not classified as disease are also involved in ADL limitation. Thus, resources supplied by others must be employed to varying degrees for many of the elderly, particularly the old-old, to carry out the activities of daily living.

NEED FOR BIOMEDICAL BREAKTHROUGHS

In addition to relying on their personal savings and employer-related pensions, the elderly have relied on governmental programs and nongovernmental sources, primarily family and friends, to meet their needs for medical services and help in daily living. Demographic changes projected for the 21st century, in particular the marked increase in the fraction of the population that is elderly and the low fraction of those in the 20 to 65 age range, are likely to markedly reduce the latter two resources, family and friends.

For example, currently in the United States, family members provide more than 80% of the help needed by the elderly for ADL and IADL. However, the increasing number of never-marrieds and childless couples and the small families of the "Baby Boomers" will mean a markedly reduced pool of family caregivers in the future. The impact will be greatest for elderly women because of the higher prevalence of nonfatal chronic diseases in women than in men; the longer life expectancy of women; and the likelihood of widowhood, not only because women have a greater life expectancy than men, but also because husbands in the "boomer" generation are usually older than their wives. Currently, about 75% of elderly men and 40%

of elderly women live with their spouse, and although this may change somewhat in the 21st century, the basic pattern is likely to remain the same.

The governments of the developed nations are beginning to have difficulties financing their support programs for the elderly, and the projected increases in the fraction of the population that will be elderly in the 21st century can only worsen the situation. Markedly increasing the fraction of wealth that would need to be transferred from the young to the old does not seem a politically feasible course of action. Of course, a modest amount of adjustment of government programs aimed at meeting the most pressing needs is possible. For example, recognizing that calendar age and biological age are different and providing benefits based on biological age may help; indeed, many over 65 years of age can continue in the work force. Of course, to use such an approach, in a manner that is fair, will require that gerontologists develop better means of measuring biological age than are currently available.

When discussing ways to meet this societal challenge, many sociologists and others concerned with public policy seem to lose sight of the fact that aging is fundamentally a biological process. For example, some public policy professionals believe that the cost of meeting the needs of the elderly will automatically be lessened if behavioral changes are made, such as stopping smoking and reducing animal fat consumption. The possibility that these behavioral changes may not decrease, but rather increase, costs does not seem to have occurred to the proponents of this view. Although it is clear that age-associated diseases, such as cancers and heart disease, can be delayed or prevented by these behavioral changes, one must consider the possibility that in the long-term these changes could increase costs. The reason for this apparent paradox is that in the absence of these fatal diseases, senescence will continue; and at more advanced ages, deterioration, including that due to nonfatal chronic disease, could result in a more prolonged period of dependency than prior to the behavioral change. If such were to be the case, then the cost to society could well be greater than before the change in behavior. Indeed, biological issues and approaches are rarely given the thoughtful consideration needed by those involved in public policy. In my opinion, the magnitude of the problem to be encountered in the 21st century because of an increasing population of elderly is very great; what is needed is the development of biomedical approaches aimed at lessening those aspects of aging that make the support of the elderly so costly. One such approach lies in the prevention or amelioration of the debilitating aspects of chronic diseases, particularly the nonfatal diseases, that afflict the elderly. Another is the development of procedures that broadly reduce age-associated deterioration, i.e., measures that decrease the extent of senescence. Both approaches have been the subject of research, much of it involving animal models, in laboratories throughout the world. Promising leads have emerged and are the subject of Chapter 7.

ADDITIONAL READING

Butler, R. N., & Brody, J. A. (Eds.) (1995). *Delaying the onset of late-life dysfunction.* New York: Springer.

Estes, C. L., Linkins, K. W., & Binney, E. A. (1995). The political economy of aging. In R. H. Binstock and L. K. George (Eds.), *Handbook of aging and the social sciences* (4th ed., pp. 346–361). San Diego: Academic Press.

Kahana, E., Biegel, D. E., & Wykle, M. (1994). *Family caregiving across the lifespan.* Thousand Oaks, CA: Sage.

Laslett, P. (1997). Interpreting demographic changes. *Philosophical Transactions of the Royal Society, London,* B, 352, 1805–1809.

Matthiessen, P. C. (1996). Demography: Impact of an expanding elderly population. In P. Holm-Pedersen and H. Loe (Eds.), *Textbook of geriatric dentistry* (2nd ed., pp. 505–516). Copenhagen: Munksgaard.

Myers, G. C. (1990). Demography of aging. In R. H. Binstock and L. K. George (Eds.), *Handbook of aging and the social sciences* (3rd ed., pp. 19–44). San Diego: Academic Press.

Riley, M. W., (1996). Age stratification. In J. E. Birren (Ed.), *Encyclopedia of gerontology* (Vol. 1, pp. 81–92). San Diego: Academic Press.

Schneider, E. L. (1999). Aging in the third millennium. *Science, 283,* 796–797.

Schultz, J. H. (1995). *The economics of aging* (6th ed.). Westport, CT: Auburn House.

Uhlenberg, P., & Minor, S. (1995). Life course and aging: A cohort perspective. In R. H. Binstock and L. K. George (Eds.), *Handbook of aging and the social sciences* (4th ed., pp. 208–228). San Diego: Academic Press.

Vaupel, J. W. (1997). The remarkable improvements in survival at older ages. *Philosophical Transactions of the Royal Society, London,* B, 352, 1799–1804.

Vaupel, J. W., Carey, J. R., Christensen, K., Johnson, T. E., Yashin, A. L., Holm, N. V., Iachine, I. A., Kannisto, V., Khazaeli, A. A., Liedo, P., Lango, V. D., Zeng, Y., Manton, K. G., & Curtsinger, J. W. (1998). Biodemographic trajectories of longevity. *Science, 280,* 855–860.

3
Why Aging Occurs

It has long been a puzzle why a phenomenon as detrimental to the individual as aging has not been eliminated by evolutionary forces but rather occurs almost universally in animals and plants. An intellectually satisfying solution to this puzzle is provided by the evolutionary theory of aging, which has been developed over the past 50 years. However, empirical evidence in support of this theory is scant, and much further research is needed to validate (or invalidate) this provocative concept. There is an urgency to do so, because in addition to its potential for solving the puzzle of why aging occurs, the evolutionary theory of aging holds the promise of providing a framework for studies aimed at determining the proximate biological mechanisms of senescence, i.e., the biological activities that directly underlie senescence. Before discussing the evolutionary theory of aging, a brief overview of the elements of evolutionary biology is needed.

EVOLUTIONARY BIOLOGY

The idea that more complex organisms evolved from simpler forms, culminating with the human species, had its origins in Europe in the middle of the 18th century. However, in the 18th and the first half of the 19th century, most scientists did not view evolution as a legitimate science, but rather viewed it as a pseudoscience akin to phrenology and astrology. This attitude was quite justified because the theoretical constructs proposed for biological evolution had little merit, and empirical evidence in support of these theories was totally lacking.

Then, in 1858, Alfred Russel Wallace published a paper in which the concept of *natural selection* was presented as the basis of evolution. About the same time, Charles Darwin, who had independently developed a similar concept, was completing a book about it, *On the Origin Of Species*, which was published in 1859. Darwin's concept of natural selection can be summarized as follows: Individuals having any advantage, however slight, have the best chance of surviving and procreating their kind, but any variation in the least degree injurious will have the opposite effect.

Natural selection provided a logically satisfying foundation for biological evolution and, in addition, Darwin presented empirical support for the theory, both in *On the Origin of Species* and in *The Descent of Man*, which was published in 1871. The concept of Wallace, and particularly of Darwin, was the beginning of recognition of evolution as a legitimate branch of biology.

It is remarkable, indeed, that Darwin and Wallace were able to conceive of natural selection without knowledge of either Mendelian genetics or molecular genetics, subjects that play key roles in our current understanding of evolutionary biology. Gregor Mendel, a Moravian monk, reported his landmark research on genetic inheritance in garden peas in 1865, more than 5 years after the initial publications of Wallace and Darwin on natural selection. In fact, the significance of Mendel's work was not recognized by biologists until 1900, some 18 years after Darwin's death. Once recognized, Mendel's work rapidly led to the development of the "classical theory of the gene" by an American, T. H. Morgan, and his colleagues, which they presented in a book, *The Mechanisms of Mendelian Heredity*, published in 1915. It was not until 1953, when the American James Watson and the Briton Francis Crick published their landmark paper on the molecular structure of DNA (the double helix), that the stage was set for the rapid development of molecular genetics.

A full appreciation of the basic tenets of evolutionary biology requires that natural selection be considered in terms of Mendelian genetics and molecular genetics. Mendelian genetics defines genes as the functional units of heredity which carry information from one generation to the next and which, along with environmental factors, determine the structural and functional characteristics (the phenotype) of an organism. Molecular genetics has provided knowledge of the molecular structure of genes and of the molecular processes involved in the functioning of genes as determinants of the phenotype of an organism. Physically, genes are components of DNA molecules. Each DNA molecule is composed of many thousands to many millions of subunits called nucleotides; and each nucleotide contains one of the following four bases: adenine (A), cytosine (C), guanine (G), and thymine (T). Most DNA is present in the nuclei of the cells of organisms in structures called chromosomes. Each chromosome contains a single DNA molecule, and although the number of different genes in a DNA molecule is not known for certain, it is estimated to be in the order of several thousand. Each gene, in turn, is composed of several thousand nucleotide subunits, and differs from other genes both in the number of nucleotide subunits and in the sequence of the four different nucleotide bases as they are linked to each other. What is meant by sequence of the four nucleotide bases is illustrated by the following hypothetical twelve subunit example: A-A-T-G-A-T-C-A-A-G-G-A. Clearly, the number of each of the nucleotide bases could be varied in this example (e.g., G could vary in number from 0 to 12) and the sequence of these four kinds of nucleotide bases could also be varied in many different ways. Since each gene contains thousands of nucleotide subunits and genes vary in total number of nucleotide bases, it is clear that the number of potentially different genes is almost limitless.

Changes in the sequence or number or both of the nucleotide bases of a particular gene do occur, and such changes are referred to as mutations. If the change in the sequence of a gene is small, the mutation may result in an altered version of the gene, called an allele. Although the frequency of mutations is estimated at about one per billion nucleotides per generation, given the thousands of nucleotides per gene, the hundred thousand or so genes per person and the trillions of cells per person, a very conservative estimate is that many thousands of mutations must occur during a human lifetime. Of course, most occur in somatic cells (i.e., cells other than ova, sperm, and their precursor cells) and thus are not passed on to progeny. Mutations occur for a variety of reasons: chemical instability of the bases of the nucleotides, errors in replicating the DNA molecule during cell division, and exposure to environmental factors such as certain naturally occurring chemicals, industrial chemicals, and electromagnetic radiation. When a mutation occurs in the gametes (i.e., the sperm cells of males or the ova of females), it can be passed to progeny and thereby can influence the evolutionary fitness of progeny.

When passed to progeny, many mutations in gametes have a detrimental effect. Indeed, such gametes often do not result in the production of viable progeny; and when viable progeny do result, they are less evolutionarily fit than others of the species (i.e., they generate fewer progeny). Therefore, the mutated gene or allele disappears from the species because of natural selection, a process frequently referred to as the gene being selected against. The more the mutated gene or allele reduces evolutionary fitness, the more rapidly it will be eliminated from the species.

However, many mutations in the gametes have characteristics that do not significantly alter the evolutionary fitness of progeny. These mutations are regarded as neutral in respect to natural selection; i.e., they are neither selected against nor selected for. Such mutated genes or alleles may not disappear from the population, but rather may fluctuate randomly in frequency of occurrence, a phenomenon called genetic drift. If they remain in the population, subsequent changes in environment could render these alleles either detrimental or favorable to the evolutionary fitness of the organism, and, at such time, their frequency in the population will be determined by natural selection.

Of the many mutations that occur in the gametes, only a very few increase the evolutionary fitness of the progeny. Such mutant genes are selected for and become the most frequently occurring allele of that gene in the population. Indeed, if the force of selection is strong enough (i.e., if fitness is greatly increased), the mutant allele will completely replace the ancestral gene, and in such cases, the mutant allele is said to have become fixed in the population.

Biological evolution results from the selection for genes or alleles that increase evolutionary fitness and also from genetic drift. Speciation (i.e., the divergence into different species) is the result of many small changes in genes over an extended period of time. A requirement for speciation is that there be a reproductive barrier between the population comprising the emerging species and that of the species from which it is emerging; this may be a physical barrier, such as a large body of water, or a biological barrier.

Before completing this brief discussion of evolutionary biology, the issue of "individual selection" versus "group selection" should be considered. The account just presented is a description of individual selection; i.e., it is the genes that the individual transmits to the progeny that are being selected for or against. This view is in accord with the writings of Darwin. Over the years, however, many biologists have viewed selection for as being for the good of the species, which is called "group selection." There is evidence for the occurrence of group selection; and often in elementary courses in biology, group selection in terms of being beneficial for the species is the only concept of biological evolution taught. As a result, most laymen and even many biologists hold this concept of biological evolution. However, currently most evolutionary biologists are in accord with Darwin and feel that individual selection plays a much greater role in biological evolution than does group selection. Of course, individual selection can often be good for the species, but individual selection is felt to dominate, whether or not it is of benefit to the species.

HISTORY OF EVOLUTIONARY THEORIES OF AGING

Let us now turn from a general consideration of biological evolution to the question of the evolution of aging. In 1870 Wallace proposed the first evolutionary explanation of why aging occurs. He theorized that after an individual has produced a sufficient number of surviving progeny and the ability to generate more progeny is diminished, the individual is still a consumer of resources and is thus detrimental to the progeny. Based on this line of thought, Wallace concluded that natural selection uses aging to eliminate old individuals from the population. In 1889, August Weismann, a renowned German biologist, embraced a similar evolutionary explanation of senescence, when he stated that in regulating length of life, the advantage to the species, and not to the individual, is what is important.

There are several flaws in these early views of the evolution of aging. First, they do not really address the question of why aging evolved, because they are based on the fact that senescent deterioration does occur with advancing calendar age. If senescence did not occur, there would be no point to eliminating the old to make way for the young, because the old and the young would generate progeny at the same rate and would not differ in other functional capacities. Thus these early views did not address the question of why senescence evolved, but rather provided a possible explanation of why senescence would be maintained.

A second flaw is that these views are based on the concept that senescence is required to eliminate the old to make way for the young. However, for most species in nature, senescence is not required for this purpose. Because of predators, infection, weather, and a host of other environmental hazards, most cohorts in natural populations are reduced to a very low number before senescence has appreciably occurred.

Still another flaw is that these views are based on group selection or, as stated by Weissman, advantage to the species. However, as just discussed, current evolutionary biologists believe group selection is weak compared to individual selection, and since senescence decreases the evolutionary fitness of the individual, natural selection would not be expected to favor it.

Nevertheless, many people still embrace the concept that senescence occurs for the good of the progeny and the species. For example, the surgeon Sherwin B. Nuland in his widely read book *How We Die: Reflections on Life's Final Chapter*, published in 1994, says that nature's job is to send us packing so that subsequent generations can flourish.

Thus the puzzle remains that senescence does occur widely in nature, even though it is detrimental to the individual and therefore would be expected to have been eliminated by natural selection. Fortunately, theoretical concepts have gradually developed that address this puzzle in a satisfying fashion. The current view of how senescence evolved is based on the concepts of population genetics, a branch of biology that was not born until the 20th century. By the 1920s and 1930s, two of the founders of population genetics, R. A. Fisher and J. B. S. Haldane, had developed the foundations for the currently held evolutionary theory of aging. However, in their published papers, neither dealt directly with the question of why aging occurs.

CURRENT EVOLUTIONARY THEORY OF AGING

It was in 1946 that Peter Medawar published a paper based on the principles of population genetics that laid out the elements of the current evolutionary theory of aging. In that paper, he states that the basis for the evolution of aging is the fact that the force of natural selection weakens with increasing calendar age. Medawar presented his views in a verbal, qualitative way. However, in the 1960s and 1970s, W. D. Hamilton and then Brian Charlesworth further advanced the concept by developing mathematically explicit procedures for assessing the weakening of the force of natural selection with increasing adult calendar age.

Based on the concepts of Medawar, Hamilton, and Charlesworth, the current evolutionary theory provides a compelling framework for why senescence evolved. Possibly the clearest way of explaining this theory is to consider a way in which senescence could develop in a hypothetical species that had hitherto been free of senescence. The premise is that this species, like most species in nature, lives in an environment that is hazardous due to predators, infectious agents, and other hostile environmental factors. Because of these hazards, the number of individuals in a cohort of this species will progressively decrease with advancing calendar age, even though senescence is not occurring in this hypothetical species. Indeed, the more hostile the environment, the more rapid will be the decrease in the size of the cohort with increasing calendar age. It is this progressive decrease in the size of the cohort that underlies the decrease in the force of natural selection with advancing calendar age. This can be visualized as follows: At the age at which the cohort reaches sexual maturity, there are a large number of individuals available to produce progeny and, as a result, this age group contributes many progeny to the next generation. With advancing calendar age, however, the number of individuals in the cohort decreases because of death due to environmental hazards and, as a result, the rate of generation of progeny by the members of this cohort decreases and becomes negligible at advanced ages when few, if any, members of the cohort are still alive. If a deleterious mutation were to be expressed early in life in some members of the cohort, it would decrease the ability of those with this mutation to generate progeny, and thus this mutation would gradually disappear in succeeding generations. However, if such a mutation were to be expressed only late in life (i.e., at ages where few if any of the cohort were still alive because of death due to environmental hazards), the mutation would have little or no effect on the number of progeny generated by those with this mutation and thus it would be passed to succeeding generations as effectively as the nonmutated gene. Moreover, under these environmental circumstances, there would be little or no ill effects due to this deleterious mutation, because it would either not be expressed or be expressed in a negligible number of the members of a cohort; i.e., all or nearly all would be dead before it is expressed. However, if a cohort of this species with this deleterious mutation were to be transferred to a protected environment, where most of the population would live to the advanced age at which the mutation is expressed, the deleterious effect would become manifest late in life, resulting in the deterioration of the organism, a condition we call senescence.

This hypothetical example illustrates how senescence could develop in a species that previously had not undergone senescence. However, the evolution of senescence in nature is undoubtedly far more complex and probably involves several different genetic mechanisms, some of which will be discussed later in this and subsequent chapters. Nevertheless, the basic principles of the current theory of the evolution of senescence are clearly illustrated by this hypothetical example.

The general concept just presented for a hypothetical nonaging species also applies to species that undergo senescence. However, in the latter, two additional factors underlie the decrease in the force of natural selection with advancing calendar age. Because of senescence, the vulnerability of the organism to environmental hazards is increased with increasing adult calendar age. In addition, unlike the members of a hypothetical nonaging population, those undergoing senescence have a decrease in fecundity with increasing calendar age. For both of these reasons, under identical environmental conditions, the decrease in the force of natural selection with advancing calendar age will be even greater in a population that undergoes senescence than in a hypothetical nonaging population.

Although humans evolved under hazardous conditions and their aging characteristics relate to the circumstances under which they evolved, most human populations are now living under conditions that greatly protect them from environmental hazards. As a result, most humans live to advanced ages and exhibit advanced stages of senescence, which in many cases reach the point where it is appropriate to use the term senility. Animals that humans protect, such as pets, also live to calendar ages at which advanced stages of senescence occur. This is not to imply that senescence does not occur in organisms living in the wild, because even in the wild some of the population of many species live long enough for aspects of the aging phenotype to appear. However, the fraction of the population of most species in the wild reaching extreme old age with a fully developed aging phenotype (i.e., senility) is vanishingly small.

GENETIC MECHANISMS IN THE EVOLUTION OF AGING

The evolutionary theory of aging does not provide specific information on genetic or physiological mechanisms involved in senescence, but it does limit such mechanisms to those that are compatible with the theory. (Possible specific physiological mechanisms will be discussed in Chapter 4.) In this chapter, the two genetic mechanisms that have been the focus of the most attention will be considered.

In 1952, Medawar proposed what is called the mutation-accumulation mechanism. He recognized that many of the mutations present in gametes are deleterious to progeny, and that such mutations are usually removed from the population by natural selection. However, he reasoned that some of these mutations may not express their deleterious action until late in life, and that such mutations will not be eliminated by natural selection, because by then the force of natural selection is low to negligible. He proposed that such mutations remain in the population and are responsible for senescence because they will be expressed in organisms living in a lifelong protected environment. That is, aging occurs because of the accumulation of mutated genes that have deleterious effects only late in life. It is this genetic mechanism that was used to illustrate how senescence could evolve in a hypothetical nonaging species.

The other genetic mechanism that has received much attention was proposed in 1957 by G. C. Williams and is referred to as antagonistic pleiotropy. Pleiotropy denotes the fact that a gene can have more than one phenotypic effect, and antagonistic pleiotropy means that the gene can have effects that oppose each other. When considering antagonistic pleiotropy in regard to senescence, Williams postulated that there are genes that have a beneficial phenotypic effect early in life and a deleterious one late in life. Such genes would be strongly selected for because of their beneficial effects early in life when the force of natural selection is great. Moreover, even if their deleterious effects late in life were catastrophic, such genes would not be eliminated from the population because the force of natural selection is weak to negligible late in life. Williams' concept of antagonistic pleiotropy provides a way that senescence can be selected for but in an indirect fashion (i.e., because of the beneficial action of these genes early in life). Thus, insofar as antagonistic pleiotropy is involved in its occurrence, senescence is a byproduct of evolutionary forces.

These two genetic mechanisms are not mutually exclusive and, indeed, both could be components of senescence in the same individual. Moreover, they are not the only genetic mechanisms compatible with the evolutionary theory of aging that could be involved in senescence. The evidence in regard to their role in senescence and a discussion of another likely genetic mechanism will be presented in Chapter 5, "Biological Basis of Aging: A Unifying Concept." At this point, our attention will turn to a discussion of the evidence in support of the evolutionary theory of aging irrespective of specific genetic mechanisms.

EVIDENCE SUPPORTING THE EVOLUTIONARY THEORY

It is difficult, if not impossible, to directly test the validity of theories of evolutionary biology, and the evolutionary theory of aging is no exception. However, it is possible to invalidate an evolutionary theory by showing that the assumptions on which it is based are false or that predictions based on it are incorrect. On the other hand, if assumptions and predictions are tested and prove to be correct, evidence in support of the theory has been obtained, although such evidence is certainly not unequivocal proof of its validity.

A major assumption of the evolutionary theory of aging is the existence of genetic variation that influences senescence. As discussed in Chapter 1, there is abundant evidence, such as the differences in the rate of aging among different mouse strains and the heritability of longevity in humans, that validates this assumption.

A prediction of the evolutionary theory of aging is that the rate of aging will be slower in populations with higher fertility rates at older ages. This prediction has been tested in laboratory studies using the fruit fly *Drosophila melanogaster,* and the results of these studies are in accord with this prediction. An example of such studies is one carried out by Michael Rose and his colleagues. They used a population of fruit flies which they divided into two groups. One group was allowed to reproduce only at young ages (i.e., an early-reproducing group) and the other group was allowed to reproduce only at older ages (i.e., a late-reproducing group). The early-reproducing group was allowed to reproduce only at 14 days of age; the late-reproducing group was allowed to mate, but their fertilized eggs were discarded until these fruit flies were many weeks older (as old as 10 weeks of age) at which time their fertilized eggs were collected over a 2-day period and used to start the next generation. After about 10 years of this procedure (involving more than 50 generations for both groups), the influence of this manipulation on aging was assessed by determining longevity characteristics of the two groups. From the survival curves shown in Figure 3.1, it is clear that the late-reproducing group evolved a much longer lived fruit fly than the early-reproducing group. In genetic terms, this study selected for genes beneficial in late life or against those detrimental in late life or both.

Another prediction is that the rate of aging of a species or subgroup of a species should be strongly influenced by the level of hazards in the environment in which the species or subgroup evolved. That is, the rate of aging should be more rapid in those that evolved in a highly hazardous environment than in those that evolved in a less hazardous environment. This prediction has proven to be correct in several instances.

For example, species of birds of the same size as species of terrestrial (earthbound) mammals have a slower rate of aging than the mammals; even more impressive is the fact that those species of birds which do not fly or which are poor fliers exhibit the same rate of aging as earthbound mammals of comparable size. The ability to fly makes the environment much less hazardous because it reduces the threat of predation and other dangers such as flood and fire. Indeed, bats (mammals that fly) have a slower rate of aging than earthbound mammals of the same size.

Another example comes from the study of S. N. Austad and K. E. Fischer, who compared populations of opossums that live on the mainland of North and South Carolina and Georgia with those that live on Sapelo Island off the coast of Georgia. These populations have been physically separated for many generations, about 4,000 years, and those on Sapelo Island have experienced a much less hazardous environment because they faced fewer predators. The rate

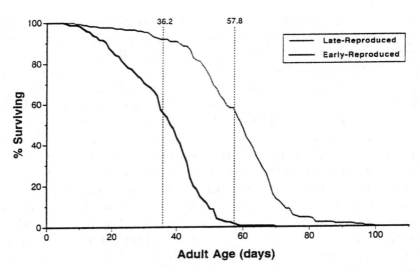

Figure 3.1 Survival curves of progeny from populations of fruit flies (*Drosophila melanogaster*) that for about 10 years were either allowed to reproduce only at a young age (early-reproduced population) or an old age (late-reproduced population).

From "Evolution and Comparative Biology," by M. R. Rose, in *Encyclopedia of Gerontology*, 1, p. 513, ed. by J. E. Birren, 1996, San Diego, CA: Academic Press. Copyright 1996 by Academic Press. Reprinted with permission.

of aging was found to be slower for the island population of opossums than for the mainland population.

Another prediction is that the rate of aging of species living in the same environment should be slower for species of large size than for those of small size because, in general, the larger the animal, the better it can cope with predators and other environmental hazards. Indeed, there tends to be an inverse correlation between the size of the members of an animal species and the rate of aging of the species.

Thus, there is some evidence to support the evolutionary theory of aging. However, much more evidence is needed to build a strong case for its validity or, of equal importance, to invalidate the theory. In particular, more studies in the wild are needed, such as that of Austad and Fischer just discussed.

Despite the need for further testing, the evolutionary theory of aging will serve as a framework for the concepts about aging to be considered in the remainder of this book. Simply put, this book is based on the assumption that the evolutionary theory of aging is valid.

ADDITIONAL READING

Charlesworth, B. (1994). *Evolution in age-structured populations* (2nd ed.). London: Cambridge University Press.
Medawar, P. B. (1952). *An unsolved problem of biology.* Oxford: Oxford University Press.
Rose, M. R. (1991). *Evolutionary biology of aging.* New York: Oxford University Press.
Rose, M. R. (1996). Evolution and comparative biology. In J. E. Birren (Ed.), *Encyclopedia of gerontology* (Vol. I, pp. 509–517). San Diego: Academic Press.
Rose, M. R. (1998). *Darwin's spectre: Evolutionary biology in the modern world.* Princeton, NJ: Princeton University Press.
Stearns, S. C. (1992). *The evolution of life histories.* Oxford: Oxford University Press.

4
How Aging Occurs

The evolutionary theory of aging has yielded important insights into why aging has evolved. It does not, however, provide information on how it occurs, that is, in terms of the specific genes, molecular events, and physiological processes that underlie the aging phenotype. In short, it does not define the proximate mechanisms responsible for senescence. However, the search for proximate mechanisms has certainly not been neglected. Indeed, it has been the major thrust of biologists interested in aging for the past hundred years, and the literature on this subject is both vast and unwieldy.

OVERVIEW

For a number of years, the premise of most gerontologists had been that there is a primary process that underlies aging, and that most characteristics of the aging phenotype result from this primary aging process, either directly or through a chain of secondary and further removed events. Indeed, most of the "theories of aging" proposed over the years were hypotheses about the nature of the primary aging process, i.e., the basic proximate mechanism responsible for senescence. Indeed, many believed that the primary aging process was the same for all species that undergo senescence. Few today feel that a single proximate mechanism is likely to underlie senescence in a particular species, or that proximate mechanisms are the same for all species.

However, there is reason to believe that in a particular species, senescence may involve a relatively small number of proximate mechanisms, and that many of these mechanisms may be common to a number of different species. In support of this view is the fact that the general pattern of aging appears to be similar for most mammalian species; e.g., with increasing calendar age, rats, dogs and humans show a loss in physiological capacity, a decrease in the ability to cope with challenges, and an increase in chronic diseases.

A consideration of a number of the "theories," hypotheses, and concepts offers a useful overview of many aspects of the aging phenotype that have been identified in a variety of species. More importantly, it provides the database needed for formulating a unifying concept of the biological basis of aging that is compatible with the evolutionary theory of aging. Such a formulation will be undertaken in Chapter 5. Quotation marks were used above with "theories of aging" and "theories" because in no case has the supporting evidence been sufficient to justify using the word "theory"; "hypothesis" would be the more appropriate term. Henceforth, however, the use of quotation marks will be discontinued; theory, hypothesis, and concept will be used, with usage based on the term used by those who proposed and championed it.

THEORIES, HYPOTHESES, AND CONCEPTS

So many theories, hypotheses, and concepts of proximate mechanisms of senescence have been proposed over the years that it is difficult to know where to begin this discussion. Probably the best approach is to categorize them by a classification system that enables an orderly presentation of the subject area. Most of the proposals can be grouped within one of the following four categories:

1. evolutionarily adaptive aging clocks
2. wear and tear
3. genes and gene expression
4. regulation of systemic function

Some of the specific theories, hypotheses, and concepts can be assigned to more than one of these four categories; in such cases, assignment to a particular category is based on my assessment of the views of those who proposed it. In addition, many of the concepts are interlinked, and such linkages will be pointed out.

Evolutionarily Adaptive Aging Clocks

Over the years, many biologists believed that aging is programmed by evolutionarily adaptive genetic clocks in a fashion similar to the programming of the development of the organism from conception to young adulthood. This notion has been popularized by articles in newspapers and magazines. However, developmental clocks are based on a model of sequential gene action which has been selected for in an evolutionarily adaptive fashion. If the evolutionary theory of aging presented in Chapter 3 is correct, senescence is not the result of adaptive selection, but rather, at most, is a byproduct of evolution. Therefore, the view of aging clocks as akin to developmental clocks is not compatible with currently held concepts of the evolutionary biology of aging.

However, one should not assume that genetic clocklike mechanisms cannot be involved in aging, nor that there is no relationship between development and aging. Senescent deterioration during adult life is clearly influenced by genetic characteristics. Although modifiable by environment, the genetic makeup of the organism is one factor influencing the rate of aging; thus, in a very limited way, senescence can be viewed as programmed by a genetic clock. As for a tie between development and senescence, one potential mechanism immediately comes to mind. Developmental processes that increase early-life fitness are strongly favored by natural selection, even if those processes are detrimental in late life, thereby becoming a cause of senescence; i.e., the genetic process of antagonistic pleiotropy could link development and senescence in a clocklike manner. In addition, in a general sense, developmental processes play an important role in determining the vulnerability of the fully developed adult to damaging challenges, including those involved in senescence. Thus, there clearly are links between development and senescence.

In conclusion, it seems warranted to discard the concept of evolutionarily adaptive aging clocks *per se*. However, some aspects of this concept may well provide useful insights in the quest for an understanding of the proximate mechanisms underlying aging.

Wear and Tear

Among the earliest theories of aging are those based on wear and tear. These theories focused on the wearing out of gross structures, such as the erosion of teeth of herbivores, which is, indeed, an important factor in the senescence of such species. The proponents of these early wear and tear theories likened the senescence of living organisms to deterioration with time of inanimate objects such as automobiles. This analogy is not totally appropriate because unlike inanimate objects, living systems utilize external matter and energy to repair wear and tear. For example, an elephant develops a new set of molars at six separate times at about 10-year intervals during its life, and thus for about 70 years, it replaces molars as they wear out. Nevertheless, with advancing age, repair processes fail to keep up with wear and tear (e.g., the elephant does not develop a seventh set of molars) and according to wear and tear concepts, it is because of this inability to completely repair wear and tear that senescent deterioration occurs. Most current theories in the wear-and-tear category are focused on cellular and molecular deterioration, rather than directly on deterioration of gross structures. Thus before discussing specific wear-and-tear concepts, it is necessary to digress and discuss some elements of cell biology.

Mammals, including humans, are made up of many different kinds of cells, which vary greatly in their structure and function (e.g., nerve cells, muscle cells, red blood cells). There are about 60 trillion cells in the human body. Figure 4.1 presents a schematic drawing of a highly simplified cell, illustrating its basic elements. Most cells contain a nucleus in which the chromosomes reside. Cells are separated from the surrounding watery medium (which in animals is referred to as the extracellular fluid) by a membrane called the cell membrane. Like all biological membranes, the cell membrane is composed of many lipid molecules (fat-like molecules) in a continuous sheet that is two molecules of lipid thick. This lipid membrane serves as a barrier to the entrance of molecules into a cell and their exit from a cell. However, embedded in the cell membrane are protein molecules that have a variety of functions, including serving as specialized pathways for the entrance and exit of specific molecules. All of the material in the cell, other than the nucleus, is collectively referred to as the cytoplasm. The cytoplasm, which surrounds the nucleus, contains a variety of different membranous

How Aging Occurs

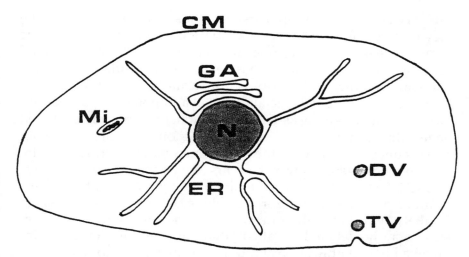

Figure 4.1 Schematic drawing of a highly simplified animal cell. Each line refers to a membrane. CM denotes the cell membrane; DV, a digestion vesicle; ER, the endoplasmic reticulum; GA, the Golgi apparatus; Mi, a mitochondrion; N, the nucleus; and TV, a transport vesicle.

From "The Design of Life," (p. 5), by R. Dulbecco, 1987, New Haven, CT: Yale University Press. Copyright 1987 by Yale University Press. Reprinted with permission.

organelles that have a variety of different structures and functions; these include mitochondria, endoplasmic reticulum, Golgi apparatus, transport vesicles, and digestion vesicles, which are illustrated in Figure 4.1. Other organelles will be described as they become relevant to the discussion of senescence. The illustration in Figure 4.1 is also simplified in that it does not indicate the number of specific organelles present in a cell; e.g., each cell contains hundreds of mitochondria, rather than just one.

Rate of Living Theory

In a book published in 1928, Raymond Pearl formally presented the Rate of Living Theory of Aging. It proposed that organisms are born with a specific amount of "vital principal" that is irreplaceable and is depleted at a rate proportional to the rate of energy expenditure (i.e., the rate of fuel use, usually referred to as the metabolic rate), and that the depletion of this substance is responsible for senescence, culminating in death. This theory is classified in the wear-and-tear category because the disappearance of this "vital principal" can be viewed as being akin to wearing out. However, the gerontologic importance of this theory does not relate to the "vital principal" aspect, since such a substance has never been identified, nor does it likely exist; rather, this theory is important because it focused on energy expenditure (i.e., metabolic rate) as a potentially important factor in aging. As early as 1908, Max Rubner had postulated, based on less than convincing evidence, that the longevity of mammalian species inversely relates to the rate of energy expenditure per kilogram of body weight. Over the years, the importance of metabolic rate as a factor influencing the rate of aging has been the subject of much research and vigorous debate. At this time, it appears unlikely that metabolic rate *per se* plays a major role in aging. However, the debate has called attention to cellular mechanisms involved in fuel utilization and how they may relate to senescence. Indeed, this subject is the focus of much current investigation and has given rise to cutting-edge cellular research on a new view of the wear-and-tear concept.

Free Radical Theory

Before turning our attention to this new view of the relationship between aging and the cellular processes involved in fuel utilization, it is useful to consider the Free Radical Theory of Aging. Proposed in 1954 by Denham Harman, this theory has an important bearing on the way fuel utilization may cause cellular damage. Free radicals are highly reactive chemical entities, containing unpaired outer orbital electrons. They are present in living organisms only at low concentrations because of their high chemical reactivity. This reactivity can cause them to modify other molecules in a damaging fashion. Harman reasoned that alterations in the structure of DNA molecules due to free radical attack could be a cause of aging. He probably arrived at this idea because of his familiarity with radiation biology; free radicals are formed by the actions of radioactive decay, X-rays, and even the ultraviolet radiation of sunlight. However, it is now recognized that although present in organisms at low concentrations, free radicals are continuously being generated by a variety of important cellular processes, and among these is fuel utilization. While for many years the studies of Harman and others did not yield strong evidence in support of this theory of aging, in recent years the concept has enjoyed a strong revival with the recognition that free radicals generated during fuel utilization play a major role in oxidative stress.

Oxidative Stress Hypothesis

Animal species as well as many other species are aerobic organisms; i.e., they require a continuous supply of molecular oxygen. In such organisms, molecular oxygen is utilized for oxidative metabolism, which efficiently converts much of the energy in the fuel components of food (i.e., the carbohydrate, fat, and protein) into ATP, a chemical form containing energy that can be readily used for physiological activities. Carbohydrates and fats have molecular structures composed entirely of carbon, hydrogen, and oxygen atoms, while protein is primarily composed of these atoms; oxidative metabolism involves a series of chemical steps which, in addition to generating ATP, yield carbon dioxide and water as the primary end-products of the reactions between the dietary fuels and molecular oxygen. Most of this oxidative metabolism and ATP generation takes place in the mitochondria (often called the powerhouses of the cell) which contain the enzymes (proteins that serve as biological catalysts) and the membrane structure to carry out these complex reactions.

If carbon dioxide, water, and ATP were its only products, oxidative metabolism would be innocuous. However, oxidative metabolism also generates small quantities of superoxide and hydroxyl radicals as well as hydrogen peroxide (collectively referred to as reactive oxygen compounds) which have the potential to damage other molecules. The superoxide and hydroxyl radicals are free radicals, and further reactions of these reactive oxygen compounds generate an array of additional reactive oxygen molecules, most of which are also free radicals. It has been estimated that some 2 to 3% of the molecular oxygen used by aerobic organisms generates superoxide radicals and hydrogen peroxide, but accurate measurement in living organisms has yet to be achieved. The Oxidative Stress Hypothesis proposes that the reactive oxygen compounds which are inevitably produced during the life of aerobic organisms cause the molecular damage that underlies senescence, a concept that relates to both the Free Radical Theory and the Rate of Living Theory of Aging.

It is well established that reactive oxygen compounds can damage biological molecules. They can cause a variety of damaging

effects on DNA, including altering the nucleotide bases and thus altering the structure and function of genes. Reactive oxygen compounds can also cause carbonylation of proteins and loss of sulfhydryl groups, thereby altering enzymatic and other functional activities of proteins as well as increasing susceptibility of these molecules to proteolysis (chemical breakdown) and thus loss of function. The structure and function of cellular membranes can also be damaged by reactive oxygen molecules because they peroxidize the polyunsaturated fatty acid components of membrane lipids.

Indeed, aerobic organisms would rapidly succumb to the damaging actions of reactive oxygen compounds were it not for the organism's systems of defense and repair mechanisms. The defense systems include enzymes, which catalytically destroy reactive oxygen compounds, and a variety of antioxidant compounds such as vitamins A, C, and E. Moreover, if damage does occur, the organism has many systems for its removal and repair, including several DNA repair processes, and systems that can degrade and replace damaged proteins as well as damaged lipids.

The Oxidative Stress Hypothesis proposes that senescence arises because the protective and repair systems are not able to totally prevent the occurrence and accumulation of oxidative damage with advancing adult calendar age. If such were the case, senescence would result from wear and tear at the cellular and molecular level, secondary to the processes involved in converting energy in food into a physiologically usable form.

There is some evidence in support of this hypothesis. Molecular oxidative damage in cellular structures has been found to increase with increasing calendar age. Also, genetic engineering has been used to generate transgenic flies that overproduce protective antioxidative enzymes, and these flies accumulate oxidative damage less rapidly and live longer than flies that have not undergone this genetic manipulation. In addition, reduction of food intake in rats and mice has been found to decrease cellular oxidative damage and to increase life span.

However, to provide solid support for this hypothesis, the following questions remain to be answered: Does the fraction of oxygen utilization diverted to the generation of damaging reactive oxygen compounds increase with increasing calendar age? Does it differ among species with different life spans? Do protective enzymes and antioxidative compounds lessen with calendar age and vary among species with different life spans? Finally, do repair processes change with calendar age and vary among species with different life spans?

Glycation and Glycoxidation Hypotheses

The Glycation Hypothesis of Aging, proposed in 1985 by Anthony Cerami, also views senescence as a consequence of fuel use. However, in this case, the focus is specifically on carbohydrate fuel; the dietary sources of carbohydrate fuel (starch, table sugar, and milk sugar) are converted by digestive process to glucose and to a lesser extent to fructose and galactose. It is these products of digestion (collectively referred to as reducing sugars) that are absorbed into the blood and used by the body as fuel.

Glycation refers to the nonenzymatic reactions of the carbonyl groups of reducing sugars with amino groups of macromolecules (large molecules) such as proteins and DNA. The chemical pathway of glycation envisioned by Cerami and his colleagues is depicted in Figure 4.2. The first step is the spontaneous formation of a Schiff base between the carbonyl group of a reducing sugar and the amino group of a macromolecule. The Schiff base can undergo an intramolecular rearrangement to form the more stable Amadori product. Over a period of several weeks, Amadori products can undergo a series of spontaneous reactions, including inter- and intramolecular rearrangements, dehydrations, and intermolecular condensations, to yield a heterogeneous array of substances collectively referred to as advanced glycation end-products (AGE), an altogether tempting acronym. Although the extent of accumulation of AGE is influenced by many factors, physiologically the most important are the concentration of the reducing sugar and the turnover of the macromolecule (its rate of degradation and replacement). For example, individuals with diabetes mellitus accumulate more AGE than do other individuals because their body fluids have higher concentrations of glucose. Also, proteins of the lens of the eye, which turn over slowly, exhibit more AGE formation than do proteins that turn over rapidly.

The Glycation Hypothesis proposes that the formation of AGE on biological macromolecules alters their structural and functional properties in a fashion that causes senescence. Evidence cited in support of this hypothesis is based in part on the concept that diabetes

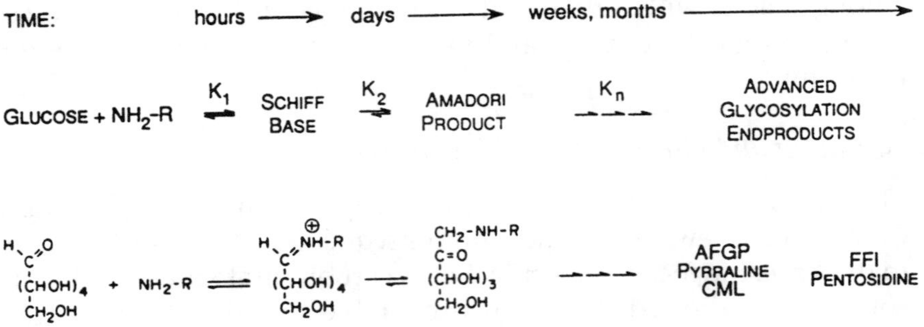

Figure 4.2 Formation of advanced glycation end products (AGE).

$$\text{TIME:} \quad \text{hours} \longrightarrow \text{days} \longrightarrow \text{weeks, months} \longrightarrow$$

$$\text{GLUCOSE} + \text{NH}_2\text{-R} \xrightleftharpoons{K_1} \text{SCHIFF BASE} \xrightleftharpoons{K_2} \text{AMADORI PRODUCT} \xrightarrow{K_n} \text{ADVANCED GLYCOSYLATION ENDPRODUCTS}$$

From "Glycation," by A. T. Lee & A. Cerami, 1996, *Encyclopedia of Gerontology*, J. E. Birren (Ed.), (Vol. I, p. 606). Copyright 1996 by Academic Press. Reprinted with permission.

mellitus is a condition of accelerated aging; and, indeed, there are similarities between the diabetic and the senescent phenotypes. Cerami and others have proposed that in diabetes and in senescence, the cross-linking of macromolecules plays an important role in the dysfunction of the organism and that AGE, which form at a more rapid rate in the diabetic, are responsible for much of the cross-linking. Cited examples of dysfunction are the cross-linking of lens proteins in cataract formation and the cross-linking of collagen resulting in increased stiffness with age of skin, arteries, and other structures. It has also been suggested that the presence of AGE on molecules of collagen (the major extracellular fibrous protein) results in the cross-linking of collagen with other proteins. This is believed to contribute to several age- and diabetes-associated problems, such as atherosclerosis, the thickening of the basement membranes of the kidney, peripheral vascular occlusions, and Alzheimer's disease. Too, there is evidence that the amino groups of the DNA bases can also undergo glycation, further opening the possibility that age changes in the structure and function of genes may result from glycation.

Recent studies have revealed a connection between glycation and oxidative stress in that an oxidative pathway can also be involved in the glycation of macromolecules, resulting in the generation of reactive oxygen compounds. Based on this information, Bruce Kristal and Byung Pal Yu proposed the Glycoxidation Hypothesis of Aging, which unites the Glycation Hypothesis and the Oxidative Stress

Hypothesis. They view wear and tear as resulting from the need of the organism to both consume oxygen for oxidative metabolism and to utilize carbohydrates as fuel.

Concept of Inadequate Protein Turnover

The proponents of this concept recognize that there are many sources of damage to the proteins of living organisms, such as the oxidation and glycation of these molecules just discussed. However, they feel that the primary cause of aging is not such inevitable damage, but rather the inadequate ability to get rid of old protein molecules and replace them with new. Indeed, it has long been known that proteins are continuously being broken down and regenerated by biosynthetic processes. Moreover, proteins are altered not only by oxidation and glycation but also by many other factors, including changes in their three-dimensional structures that appear to be due solely to the length of time they spend in cells and body fluids. In addition, there is evidence that with increasing calendar age, many species of proteins show a slowing in the rate of turnover. This concept postulates that the decrease in the rate of protein turnover is a proximate contributor to aging by allowing levels of altered, malfunctioning protein molecules to accumulate in the organism, consequently decreasing the level of fully functional molecules. The fact that essentially all physiological processes are dependent on appropriate functioning of a variety of protein species gives some credence to the view that inadequate protein turnover could, at least in part, underlie senescence.

Concept of Altered Biological Membranes

As briefly discussed earlier, biological membranes play an important role in the functioning of cells, serving as a selective barrier between the contents of the cell and the extracellular fluid, and as a site of information input from the nervous system, endocrine system, and other sources. Membranes also play a key role in the functioning of cellular organelles, such as in the generation of ATP by mitochondria. These functions depend not only on the barrier property of the membrane lipids but also on their fluid characteristics, which give the embedded proteins mobility within the membrane structure. The fluidity of membranes is influenced by two factors: the nature of the fatty acid components of the membrane and the amount of cholesterol in the membrane. The more unsaturated the fatty acid components, the more fluid is the membrane; and the more cholesterol in the membrane, the less fluid it is.

In 1978, Zs-Nagy proposed the Membrane Hypothesis of Aging in which the deterioration of cellular membranes was viewed as the major factor underlying senescence. Biological membranes do, indeed, become less fluid with advancing calendar age. In part, this may be due to oxidative stress, because it has been shown that peroxidative damage of membrane lipids does occur with increasing calendar age. In addition, and possibly more importantly, the ratio of saturated to unsaturated fatty acids in the membrane is known to increase with increasing age. However, while there is some evidence for changes in membrane function with advancing age, the role of these changes in senescent deterioration has not yet been established.

Concept of Altered Extracellular Matrix

The extracellular matrix, which surrounds the cells, markedly influences cellular functions such as cellular proliferation and migration, the response to cytokines (locally secreted small proteins that influence cell function), and gene expression. In addition, the role of the extracellular matrix transcends that of the individual cells in that it plays an important part in the structure and function of the tissues and organs of the organism. Thus, age changes in the extracellular matrix have also been viewed as a potentially major factor underlying senescence.

The extracellular matrix is composed primarily of fibrous structural proteins and complex carbohydrate polymers that bind large amounts of water to form a gel-like structure. The major fibrous structural proteins are collagen and elastin, both of which turn over very slowly.

The characteristics of the extracellular matrix, and in particular collagen and elastin, change throughout the life span of the organism. During the developmental period of life, enzymatically mediated cross-linking of these fibrous proteins results in the mature extracellular matrix of the young adult. However, because mature collagen and elastin molecules turn over very slowly, oxidation, glycation, and other nonenzymatic events gradually change the functional characteristics of these structural proteins by forming complex cross-links and other structural alterations with advancing calendar age. These postmaturational changes alter the physical properties of the organs and tissues (e.g., lessening distensibility and mechanical strength) so as to adversely affect the functioning of the cells that reside in the matrix as well as the gross functional structure of the tissues and organs.

Clearly, there are cogent reasons to seriously consider postmaturational alterations of extracellular matrix to be involved in senescence. Of course, both the oxidative stress and the glycation hypotheses are closely related to this concept of altered extracellular matrix, but other mechanisms may also be involved in the age changes in extracellular matrix. The question that plagues most attempts to assess the basic proximate mechanisms underlying senescence also arises here in that it is difficult to know to what extent changes in the extracellular matrix cause senescence or are a result of senescence. In this case, the answer would appear to be both.

Genes and Gene Expression

The theory that senescence is the result of genetically programmed evolutionarily adaptive clocks has lost favor in recent years because it is not in accord with the currently held evolutionary theory of aging. However, there are other genetic theories of aging compatible with the evolutionary theory, and these warrant our consideration. Two of these were presented in Chapter 3: the accumulation of mutated genes with deleterious actions occurring late in life; and antagonistic pleiotropic gene expression. There is no need to further discuss these theories here other than to emphasize two important aspects common to both. First, both involve the passage of genetic traits by the gametes of parents to progeny, and thus the relevant gene or genes is/are present in all of the progeny's cells that have nuclei; in the human, this is in the range of 60 trillion cells. Second, both theories are compatible with what would appear to be the involvement of a genetic clock in senescence—not one that has been selected for to produce senescence, but rather genetic mechanisms with detrimental effects in late life that have been selected for because of beneficial effects in the young adult or that have not been selected against because their effects in the young adult are neutral. Such mechanisms could well have the appearance of a genetic clock.

A great number of other theories, hypotheses, and concepts have been proposed over the years regarding the role of altered genes and deranged gene expression in senescence. In the case of many of these, it would be just as appropriate to classify them in the Wear-and-Tear category as in the Genes and Gene Expression category. Several of these concepts are worth presenting here.

Before doing so, however, it is necessary to digress briefly to describe the process of gene expression, i.e., the mechanisms by which the DNA-coded information is a determinant of the organismic phenotype. Two processes are sequentially involved: transcription and translation. Transcription occurs in the nucleus of the cell and utilizes the DNA of a gene as a template for the generation of RNA (ribonucleic acid) molecules with a sequence of four different nucleotide bases that are determined by the sequence of nucleotide bases of the gene being transcribed. (As can be imagined, this process involves a complex series of enzymatic chemical reactions; however, they need not be considered for an understanding of the elements of gene expression.) Once formed, the RNA leaves the cell nucleus and enters the cytoplasm where the process of translation takes place. Translation refers to the reactions by which the sequence of nucleotide bases in an RNA molecule codes for the formation of a particular protein containing a specific number and sequence of 20 different kind of amino acids. The number, kind, and sequence of the amino acids in a protein determine its structure and function. Thus, each gene gives rise by the sequential processes of transcription and translation to a specific protein referred to as a protein species. Humans have some 60,000 different protein species, and each protein species has one or more functions, such as enzymatic catalysis of a specific chemical reaction, involvement as a component of the contractile machinery of muscle or as a component of the structural support system of an organ, and many others. In summary, genetics influences the phenotype of an organism by directing the generation of protein species with specific functional and/or structural characteristics.

Somatic Mutation Theory

In 1959, Leo Szilard, a distinguished atomic physicist, proposed the Somatic Mutation Theory of Aging, a theory that probably arose from his interest in the biological effects of radiation and from the fact that even sub-lethal radiation almost always shortens the life span of mammals. Szilard's theory should not be confused with

Medawar's concept of the accumulation of mutated genes whose deleterious consequences are expressed only in later life. The Medawar concept refers to inherited genes, which are present in all the nucleated cells throughout the life span of the organism. In contrast, Szilard's concept is based on the generation of mutations in individual somatic cells during the life span of the organism. He proposed that genes in the chromosomes of somatic cells are inactivated by random "aging hits." Since Szilard was an atomic physicist, he was probably thinking of a gene being "hit" by radiation. He further postulated that these random hits lead to dysfunctional cells and to the death of the organism when a sufficient number of cells become dysfunctional.

This theory spurred much research, but little evidence in its support. What this research did yield is an understanding of what happens to DNA in the chromosomes of somatic cells during the lifetime of an organism. Chromosomal DNA is continuously being damaged by a variety of causes, both endogenous factors (e.g., body heat, oxidative stress, glycation, alkylation, steroid hormones, and excitatory amino acids) and exogenous factors (e.g., ionizing radiation, smoking, sunbathing, and chemicals generated by barbecues). It has been estimated that in human cells, damage to DNA from oxidative stress alone occurs at the rate of 10,000 "hits" per cell per day. Fortunately, organisms have several systems that repair most of the DNA damage. However, repairs do not quite keep up with damage and, as a result, there is an increase in the level of chromosomal DNA damage with increasing calendar age (i.e., DNA suffers from wear and tear). Thus the question to be answered is not whether damage to DNA increases with age, but rather if enough somatic cells are sufficiently damaged to cause significant organismic dysfunction. Currently available information does not answer the question of the extent to which the functional ability of somatic cells is compromised by chromosomal DNA damage resulting in mutations.

Moreover, damaged cells, including those with damaged DNA, are eliminated by apoptosis (programmed cell death). By eliminating damaged cells, apoptosis may prevent functional disturbances due to abnormal activities of damaged cells. There is a decrease in apoptosis with advancing age and by allowing damaged cells to accumulate in the organism, this decrease in apoptosis could be a cause of senescence. On the other hand, a loss of even damaged cells could be a cause of senescence and if so, the decrease in apoptosis with advancing age could be a factor retarding senescence. Thus, at this time, it is not known whether the age-associated decrease in apoptosis promotes or retards senescence.

Although the currently available information does not enable us to assess the importance of somatic mutations in senescence, there is one aspect of the aging phenotype that may well be the result of somatic mutations, and that is the age-associated increase in the occurrence of cancers. In the case of colon cancer, it has been found that several mutations occurring over a period of time are required for a cell of the colon to become cancerous. It seems likely that similar events are involved in the genesis of other kinds of cancer. For this reason alone, the Somatic Mutation Theory of Aging warrants further study.

Concept of Mitochondrial DNA Damage

Although the DNA of the chromosomes contains almost all the genes of a cell, a very small number of genes reside in the DNA of the mitochondria, specifically some of the genes that code for proteins involved in the oxidative processes that occur in the mitochondria.

Of course, mitochondrial DNA is in close proximity to the mitochondrial sites of reactive oxygen molecule generation, unlike the chromosomal DNA that resides in the nucleus of the cell. Thus it might be expected that mitochondrial DNA would undergo more oxidative damage than chromosomal DNA. Moreover, the nucleus has well-developed DNA repair systems, while such systems in the mitochondria appear to be less effective. Indeed, it has been shown that mitochondrial DNA accumulates more damage than does chromosomal DNA. The highest levels of damage to the mitochondrial DNA are found in the cells of highly oxidative organs in which little or no cell division occurs, such as heart, brain, and skeletal muscle. It has been proposed that damage to mitochondrial DNA plays a major role in senescence by interfering with the ability of the cells to generate sufficient amounts of ATP.

Although there is strong evidence for accumulation of mitochondrial DNA damage with increasing age, the evidence pointing to a major role of altered mitochondrial DNA in senescence is weak. Even at advanced ages, only a very small percentage of the mitochondrial DNA in a tissue or organ appears to be damaged. However, this field of study is new, and some researchers suspect that improved technology may show this to be an underestimate. Moreover, it appears that the damage is not broadly distributed among the cells of a tissue, but rather is highly concentrated in particular cells. Those cells whose mitochondrial DNA is very damaged may then be eliminated by apoptosis, thereby decreasing the accumulation of damaged mitochondrial DNA in an organ but at the cost of losing cells; indeed, those tissues and organs with the greatest mitochondrial damage with advancing age are also those that tend to lose the most cells with advancing age. Clearly, further study is needed before any firm conclusion can be drawn about the role of mitochondrial DNA damage in senescence.

Telomere Senescence Theory

The functional activities of some tissues and organs require certain cells to divide and yield two daughter cells by a process called mitosis. Examples are cells of the skin, intestinal mucosa, bone marrow, and immune system. To yield two daughter cells with genetic material identical to that of the mother cell requires that the chromosomal DNA be replicated, and that each daughter cell receive one of the two copies of the chromosomal DNA. Although this requirement is almost fulfilled, there is a small loss of DNA at the ends of the linear chromosomal DNA because the process does not replicate DNA at those points. The DNA at both ends of the chromosomes are called telomeres and are composed of a specific DNA nucleotide base sequence and its associated proteins. (This loss of telomere DNA, which occurs each time a cell divides, is a clear example of wear and tear at the molecular level.) Before considering how this loss in telomere length may underlie senescence, it is necessary to briefly discuss what is meant by the "Hayflick limit."

In 1961, Hayflick and Moorhead reported that a cell type called the fibroblast divides only a finite number of times when removed from the body and placed in a cell culture system. This finding has been confirmed by many others and has also been shown to occur with many other cell types. It appears to be a property of all normal somatic cells in culture and is now known as the "Hayflick limit." Hayflick hypothesized that the phenomenon of limited cell proliferation in culture is an *in vitro* ("test tube") model of aging and therefore a vehicle for the experimental study of aging. This concept was well received and for the past 30 years has been vigorously studied because of its simplicity and the ease with which the model can be manipulated, as compared to studying complex, intact organisms such as rats or humans. Although the evidence that this *in vitro* system is pertinent to organismic aging is weak, Hayflick's hypothesis continues to be popular, and the belief that this model is relevant to organismic senescence is the basis of the Telomere Senescence Theory.

This theory proposes that the shortening of the telomeres is responsible for the limited number of mitotic divisions that somatic cells undergo in culture. Thus telomere shortening is viewed as the way the cell "counts" the number of divisions it has undergone. It is further postulated that as the telomeres shorten, gene expression is altered in such a way as to finally turn off cell division. In support of this concept, the germ cells and most tumor cells (both of which have an unlimited capacity to divide in culture) have an enzyme called telomerase which promotes the extension of telomeres, thereby countering the shortening of telomeres during mitosis.

Although this theory may provide a mechanistic basis for the "Hayflick limit" in culture, the question that must be addressed is its role in the aging of the total organism. It has been shown that somatic cell division *in vivo* (i.e., in cells residing in the organism) also results in shortening of the telomeres. Does this telomere shortening lead to inadequate cell proliferation or any other functional deficit in the organism? In this regard, it has been suggested that the shortening of telomeres leads to such aging problems as inadequate wound healing, immune dysfunction, and age-associated cardiovascular disease. However, evidence in support of its role in these or any other senescent problem is scant to the point of nonexistent. Indeed, it is equally plausible that the "Hayflick limit" of somatic cell proliferation may be a mechanism that evolved to protect organisms from developing cancers. That is to say, cells that cannot undergo cell division are not likely to give rise to the uncontrolled cell division that characterizes cancers.

Concepts Based on Altered Gene Expression

Although DNA stores the genetic information, proteins are the molecules that play a primary role in carrying out the many and varied activities of living. Thus it is not surprising that several concepts of the cause of aging have focused on alterations in the processes of gene expression (transcription and translation) which generate specific proteins. In 1961, Zhores Medvedev was the first to postulate that errors in information transfer from DNA to proteins may underlie senescence. Two years later, Leslie Orgel proposed the Error Catastrophe Theory, which played a dominant role in the thinking of biological gerontologists for the next 10 to 20 years. This theory posits that if an error is made in the formation of a protein involved in the transcription or translation processes, this defective protein (i.e., a protein with an error in its amino acid sequence) would cause the generation of other proteins with errors, including proteins involved in transcription and translation. Thus the number of error-containing proteins would expand and ultimately cause what Orgel called an "error crisis" and thus the death of the cell. This theory caught the imagination of gerontologists and engendered related theories as well as a host of studies aimed at testing the validity of such theories. Although these studies generated much valuable information, they did not provide evidence in support of the Error Catastrophe Theory or related theories. Subsequently, technical advances led to studies that assessed the extent to which proteins with altered amino acid sequences were formed with advancing age, and the conclusion is that they occur infrequently. Thus neither the Error Catastrophe Theory nor related concepts appear to be valid.

An important byproduct of the Error Catastrophe Theory is the attention it directed to the possibility that alterations in gene expression play an important role in aging. In 1982, Richard Cutler proposed The Dysdifferentiation Theory of Aging. This theory is based on the fact that during development most of the cells of the body become highly specialized (e.g., nerve cells, muscle cells, etc.) by a process called differentiation. Although these cells have a full complement of DNA and its genes, each specialized cell type expresses only the fraction of those genes required for its specialized function. The rest of the genes are said to be "turned off." The Dysdifferentiation Theory posits that with increasing age, some of those "turned off" genes are "turned on," resulting in the cell generating species of proteins that are not needed, thus disturbing the functioning of the specialized cell. This concept was tested by determining changes in the species of proteins present in tissues with increasing age and, qualitatively, little change was found. Thus it is most likely this intriguing and imaginative theory has little validity.

Although, qualitatively, the young and the old have the same species of proteins in their tissues, quantitative differences have been found in both the amount and rate of turnover of particular protein species. These findings have focused attention on the transcription of specific genes and on the process of translation. The process of transcription yields a kind of RNA called messenger RNA (mRNA). Each gene yields a specific mRNA which has a sequence of the four nucleotide bases that code for a particular protein species. The amount of the specific mRNA molecule in a cell is an important factor in determining the rate of translation and thus the rate of synthesis of the protein species. The influence of age on the amount of several specific mRNA species has been measured; with increasing age, the amount of some specific mRNA molecules has been found to increase, some to decrease and some to remain unchanged. A change in the amount of a specific mRNA could be due to a change in the rate of transcription of a gene or a change in the rate of degradation of the mRNA. In the case of the few specific mRNA molecules that have been studied in this regard, a change in the rate of transcription of the gene is the cause of the change in the amount of mRNA with increasing age. Thus, based on the limited amount of information currently available, it appears that changes in the transcription of specific genes may play a role in the proximate mechanisms underlying senescence. Indeed, the factors responsible for the age changes in the rate of transcription of specific genes have begun to be studied.

In this regard, attention has been focused on the fact that there are cellular processes by which methyl groups are bonded to DNA (called methylation) and processes by which methyl groups are removed from DNA (called demethylation). Molecular biologists have shown for at least some genes that the greater the extent of methylation, the slower the rate of transcription. Initial studies by gerontologists showed that the methylation of total cellular DNA decreases with advancing age, which suggested the possibility that the rates of transcription of many genes might increase with increasing age. However, further studies, admittedly limited in scope, have found no correlation in age changes in the rate of transcription of a gene and changes in the extent of methylation of the gene.

Recently attention has shifted to the cell proteins called transcription factors, which are involved in the initiation and regulation of transcription. These transcription factors stimulate (or repress) the transcription of particular genes by binding to specific regions of the DNA comprising the gene. Changes in amount or functioning of transcription factors could well be involved in the age-associated changes in the rate of transcription. This area of investigation is just beginning, and it is too early to assess the role of alterations in transcription factors in senescence. However, recent findings of Ahmad Heydari and his associates, on the expression of a protein called heat shock protein 70, indicate that transcription factors may play an important role. Heat shock protein 70 is expressed in response to many agents, including heat, that tend to damage cells, and this protein protects the cells from such damaging agents. In studies using the rat model, Heydari and associates found that the expression of heat shock protein 70 in response to heat stress is decreased with increasing age, and that a decrease in the rate of transcription of the heat shock protein 70 gene is responsible. They further found that the amount of heat shock 70 transcription factor does not decrease with increasing age, but rather that its ability to bind to the relevant region of the DNA of the heat shock protein 70 gene is decreased, and because of this, the rate of transcription of the gene is decreased.

In contrast to transcription, translation appears to be decreased with increasing age for all species of protein; i.e., some general mechanism slows the process of translation. However, despite this decrease in the rate of translation, the expression of some proteins increases with age because an increase in their transcription more than compensates for the decrease in translation.

It is clear that aging alters the rate of expression of many genes, and this area of research certainly warrants further study. At this time, however, it is not known whether the changes that have been noted in gene expression play a causal role in senescence or are simply the result of it.

Regulation of Systemic Function

It has been proposed that aging is not due to major changes in the functioning of most cells, but rather to a small change in the functioning of specific cells involved in the integrated systemic functioning of the organism. In fact, Robert Rosen postulates that the proximate mechanisms of senescence may not involve any change in the functioning of cells, but rather may stem from the loss of precision in organismic regulatory functions. Of course, such a loss would then cause widespread changes in the functioning of many, if not all, cells of the organism.

Most of the proposed mechanisms relate to alterations in endocrine and neuroendocrine regulation of systemic organismic function. In addition, altered immune function (another example of altered systemic function) has been proposed as a proximate mechanism underlying aging.

Endocrine and Neural Regulation

There are many different types of specialized cells in diverse locations in the body, such as the pituitary gland, adrenal glands, pancreas, and ovaries, which secrete substances called hormones into the blood. Each type of cell secretes a specific hormone, such as growth hormone, cortisol, insulin, and estradiol. Each hormone then serves as a chemical messenger and is carried by the cardiovascular system to target sites where it plays a role in regulating the function of the cells at the target site. Not only do these messengers regulate the functioning of the target cells, but they do so in a fashion that integrates the functioning of these cells with the needs of the entire organism.

The concentration of the hormone in the blood is an important factor in its regulatory function; concentrations that are too high or too low can distort the regulation of organismic function. Indeed, the idea that senescence is due to a hormone deficiency has a long history. It was first proposed by Brown-Sequard, the renowned 19th century French physiologist, who in 1889 published a paper describing his attempts to rejuvenate elderly men with extracts from the testes of guinea pigs and dogs. It is likely that any success he may have had was due to a placebo effect, since the extracts probably contained insignificant amounts of the male sex hormone testosterone.

During the first half of the 20th century, the concept that aging was due to a deficiency of thyroid hormone was held by many. This belief was based on the similarities noted between elderly people and patients with hypothyroidism (the clinical condition resulting from too little thyroid hormone). However, subsequent studies have shown that elderly people do not have markedly reduced levels of thyroid hormone, and it is now felt that the noted similarities are probably superficial and not indicative of senescence resulting from thyroid hormone deficiency.

Nevertheless, the concept that a hormone deficiency may play a major role in senescence continues to have many advocates. In this regard, much attention is currently being paid to the possible role of one or more of the following hormones: growth hormone, estrogen, melatonin, and dehydroepiandrosterone (DHEA). Each will be discussed further in Chapters 6 and 7.

In point of fact, viewing the role of a hormone *solely* in terms of its concentration in the blood is an oversimplification: The effects of a hormone relate not only to its level in the blood, but also to the level and characteristics of the hormone receptor proteins in the target cells and to intracellular events, subsequent to the hormone-receptor interaction, that transform the hormone signal into a cellular response (referred to as cellular signal transduction). Thus senescence could result from changes in the hormone level in the blood or in the hormone receptor or in any of the many sequential steps in cellular signal transduction. As will be discussed in Chapter 6, we are just beginning to study the many aspects of endocrine function in regard to senescence.

The nervous system also plays a major role in the regulation of systemic function. However, remarkably little attention has been paid to the role of altered neural regulation in systemic function as a cause of aging. Nerve cells interact with target cells (i.e., other nerve cells, muscle cells, and gland cells) by releasing chemicals called neurotransmitters. These then react with receptors in the target cell and thereby set into operation cellular signal transduction reactions. Therefore, exploring the role of the nervous system in senescence will require the same kind of detailed assessment as just discussed for the endocrine system.

The neuroendocrine system, which has been the subject of much conjecture in regard to senescence, refers to the hypothalamus and its actions on the pituitary gland and the actions of the pituitary hormones on target sites. The hypothalamus is a region of the brain which receives information about the organism's external and internal environment via nerve fibers and chemicals in the blood. It is functionally connected to the posterior lobe of the pituitary gland by nerve fibers and to the anterior lobe by a specialized blood supply into which it secretes hormones. These hormones, in turn, regulate the production of hormones by the anterior pituitary gland; the latter then secretes its hormones into the general circulation. Since the neuroendocrine system plays a major role in the regulation of systemic function, it is not surprising that it has been the subject of much thought and study in regard to senescence. Details of the functioning of the neuroendocrine system are presented in Chapter 6. However, at this juncture, it is instructive to consider a hypothesis that implicates a derangement in the functioning of the neuroendocrine system as a proximate mechanism of senescence.

An excellent example is The Glucocorticoid Cascade Hypothesis of Aging, proposed by Robert Sapolsky and his associates in 1986. Before considering this hypothesis, one needs to know the elements of the functioning of the relevant neuroendocrine system. The adrenal cortex secretes glucocorticoids into the general circulation at a rate that is controlled by information sent to it by the hypothalamic-pituitary system. Glucocorticoids play an important role in the ability of the organism to cope with both physical and psychological stress; but when present at concentrations that are too high, glucocorticoids can cause widespread damage. For that reason, the organism has a negative feedback control system that senses the plasma level of glucocorticoids and adjusts the rate of secretion of glucocorticoids accordingly, much in the way a thermostat system maintains the temperature of a home. Nerve cells in a region of the brain called the hippocampus are a component of the negative feedback system; they sense the level of plasma glucocorticoids, and if the level is too high, they send information via nerve impulses to the hypothalamic-pituitary system to decrease the rate of glucocorticoid secretion by the adrenal cortex. However, this negative feedback system can be overridden when the organism is responding to stress, and the result is periods of high levels of plasma glucocorticoids.

The Glucocorticoid Cascade Hypothesis posits that encountering stress is an inevitable fact of life, and that the resulting high levels of glucocorticoids tend to damage nerve cells in the hippocampus, particularly with the simultaneous occurrence of another insult, such as a decreased flow of blood to the hippocampus. During such a time, it is proposed that some of the hippocampal neurons die, and this compromises the effectiveness of the negative feedback system. It is further envisioned that the occurrence of many such episodes during the life span leads to an ever-increasing or "cascading" increase in the plasma glucocorticoids to very high sustained levels, which causes the widespread damage characteristic of senescence, such as impairment of the immune system, loss of bone mass, and impaired cognition.

Sapolsky and his associates have provided evidence in support of this hypothesis in studies carried out with rats. They found that the plasma glucocorticoid levels increased with the increasing age of the rat, and that there was a loss of nerve cells in the hippocampus with increasing age. This elegant hypothesis was initially widely accepted. However, subsequent rat studies by other investigators have shown that neither the magnitude of the age-associated increase in plasma glucocorticoid levels nor the time course of the increase is in accord with the Glucocorticoid Cascade Hypothesis. Moreover, findings from studies with humans, which will be presented in Chapter 6, do not support the hypothesis. Nevertheless, this hypothesis does illustrate how damage to a select group of cells could lead to the kind of general systemic dysfunction that is characteristic of senescence.

Altered Immune Function

The immune system protects an organism against invading agents, such as damaging bacteria and viruses, and also against mutant cells (such as cancer cells) that arise during the life of the organism. It does so by destroying mutant cells, foreign cells (e.g., bacteria), and cells infected with viruses. It is critical that the immune system not attack the cells of the organism it protects, which immunologists refer to as the ability to discriminate between self and non-self. With advancing age, this ability tends to become faulty and, in some people, can lead to autoimmune diseases, such as rheumatoid arthritis, systemic lupus erythematosus, and glomerulonephritis (a kidney inflammatory disease), and autoimmune hemolysis (destruction of red blood cells). It is thus not surprising that altered immune function has been the basis of a number of concepts regarding the proximate mechanisms of aging.

An early, well-thought-out concept of the involvement of an altered immune system in senescence is *The Immune Theory of Aging*, proposed in Roy Walford's book published in 1969. Although Walford does not claim that rank autoimmune disease is a general occurrence during aging, he does propose that changes occur with age because of somatic mutation or some other alteration (e.g., crosslinking of macromolecules), and that these changes result in inappropriate interactions between the cells of the immune system and other cells of the body. He further proposes that as a consequence of these interactions, the functions of cells of the body are altered in such a way as to result in senescence. Although autoimmune diseases increase with advancing age, there is no evidence for the occurrence of the more subtle loss in self-recognition suggested by Walford. Nevertheless, this systemic theory is intriguing and warrants further testing as the tools for immunologic research become more advanced.

CONCLUSIONS

Obviously there has been no lack of concepts of the basic mechanisms causing senescence. However, none of them has been supported by strong evidence. There are probably two reasons for the lack of supportive evidence. First, many of the proposed mechanisms have been put forward as the sole mechanism. Since senescence probably stems from many different proximate mechanisms and their interactions, strong evidence in support of a single mechanism seems highly unlikely. Second, most of the proposed mechanisms are based on a particular characteristic of the aging phenotype. Unfortunately, it is difficult to know whether such a characteristic plays a causal role in senescence or is rather the result of senescence. Nevertheless, I believe the research underlying many of the concepts has provided a sufficient database to permit one to present a unifying concept of the biological basis of aging. In the next chapter, the author will attempt to do so.

ADDITIONAL READING

Beckman, K. B., & Ames, B. N. (1998). Mitochondrial aging: Open questions. *Annals of the New York Academy of Sciences, 854,* 118–127.

Finch, C. E. (1990). *Longevity, senescence, and the genome.* Chicago: The University of Chicago Press.

Johnson, F. B., Sinclair, D. A., & Guarente, L. (1999). Molecular biology of aging. *Cell, 96,* 291–302.

Levine, R. L., & Stadtman, E. R. (1996). Protein modifications with aging. In E. L. Schneider & J. W. Rowe (Eds.), *Handbook of the biology of aging,* (4th ed., pp. 184–197). San Diego: Academic Press.

Ozawa, T. (1997). Functional changes in mitochondria associated with aging. *Physiological Reviews, 77,* 425–464.

Rattan, I. S. H., & Toussaint, O. (Eds.) (1996). *Molecular gerontology: Research status and strategies.* New York: Plenum.

Rubin, H. (1997). Cell aging in vivo and in vitro. *Mechanisms of Ageing and Development, 98,* 1–35.

Schöneich, C. (1999). Reactive oxygen species and biological aging: A mechanistic approach. *Experimental Gerontology, 34,* 19–34.

Van Remmen, H., Ward, W., Sabia, R. V., & Richardson, A. (1995). Gene expression and protein degradation. In E. J. Masoro (Ed.), *Handbook of physiology: Section 11, Aging* (pp. 171–234). New York: Oxford University Press.

Viidik, A., & Hofecker, G. (Eds.) (1996). *Vitality, mortality, and aging.* Vienna: Facultas-Universitätlag.

Wallace, D. C. (1997). Mitochondrial DNA in aging and disease. *Scientific American, 277*(2), 40–47.

Yates, F. E. (1996). Theories of aging: Biological. In J. E. Birren (Ed.), *Encyclopedia of gerontology,* (Vol. 2, pp. 545–555). San Diego: Academic Press.

5

Biological Basis of Aging: A Unifying Concept

Over the years, many articles have been published in scholarly journals and books on what the authors believed to be the basic nature of aging. Subsequently, in newspapers, magazines, and most recently on the TV evening news, some of these theories have been proclaimed THE cause of aging. With time, these heralded causes of aging have not withstood critical scrutiny. Although a prodigious amount of research has led to some agreement as to *why* aging occurs, *how* aging occurs has remained a puzzle. However, with the massive body of available information serving as a database, it is now possible to address this apparent enigma by formulating a general unifying concept of the biological basis of aging.

The first step is to cull the mass of ideas and data on possible proximate mechanisms underlying senescence, seeking those concepts and findings that can serve in building the foundation for a unifying concept. This task is, of course, formidable. The amount of information is vast, and it is difficult to select that which provides basic information about aging from that which merely characterizes the aging phenotype. In my opinion, the tools necessary for this effort are the evolutionary theory of aging and its possible underlying genetic mechanisms as well as the recently proposed concept of "public" and "private" mechanisms of aging.

EVOLUTIONARY THEORY AS A GUIDE

Assuming the evolutionary theory of aging is valid, concepts can be evaluated in terms of their compatibility with this theory. For example, there have been many proposals that senescence results from evolutionarily adaptive "aging clocks" (i.e., genetic programs that evolved by natural selection to produce senescence). Such concepts can be discarded because they are not compatible with the evolutionary theory of aging, which holds that senescence is not selected for, but is at most a byproduct of natural selection.

In regard to likely genetic mechanisms, processes with detrimental effects manifested only at advanced ages could well underlie senescence. Indeed, the basic premise of the evolutionary theory of aging is that senescence occurs because the force of selection against it decreases progressively with postmaturational calendar age. Medawar's proposal that senescence results from the accumulation of mutations with deleterious late-life effects is an example of this line of thought. However, limiting this concept to mutations is probably too narrow a view; it should be expanded to include the genetic makeup of the organism. The genome of an organism of a particular species or subgroup of a species is, among other things, a product of environmental hazards that confronted the species or subgroup during its evolution in the wild. There are detrimental genetic traits that are not expressed until ages at which most of the population in the wild have succumbed to environmental hazards; these traits are not eliminated by natural selection. However, such traits are likely to be contributors to senescence because they will be expressed in a protected environment that enables most of the population to live to advanced ages seldom reached during evolution in the wild. This kind of genetic trait should be carefully considered in formulating a unifying concept of senescence.

In his antagonistic pleiotropy concept, Williams proposed that senescence may be a byproduct of evolutionary forces; specifically, if genetic traits increase evolutionary fitness at young ages, such traits would be selected for, even if they have detrimental results at advanced ages. Thus, in this indirect fashion, senescence can appear to be selected for, but what is really being selected for is the beneficial early-life effect. Genetic traits with this characteristic are clearly good candidates as contributors to senescence and, therefore, must be considered in developing a unifying concept.

THE DISPOSABLE SOMA THEORY OF AGING

Proposed by Thomas Kirkwood, the Disposable Soma Theory is based on the premise that, from the standpoint of evolutionary biology, the basic function of all organisms is to utilize free energy in the environment (such as that supplied by carbohydrate, fat, and protein fuels in food) to produce progeny. To do so requires that a portion of the energy be used for the maintenance of the organism, which Kirkwood refers to as the energy used for somatic maintenance. He proposes that the force of selection acts to apportion the use of energy between reproduction and somatic maintenance in a fashion that tends to maximize evolutionary fitness (i.e., tends to maximize the life span yield of progeny of the individual organism) and that as a consequence, less energy is used for somatic maintenance than is required for indefinite survival—and thus the occurrence of senescence. Based on this concept, Kirkwood further postulates that the apportionment of energy between reproduction and somatic maintenance varies according to the environment in which the species or subgroup of a species evolved. In the case of those that evolved in a highly hazardous environment (predation, infection, etc.), the greatest yield of progeny would be achieved when relatively little energy is invested in somatic maintenance, enabling such organisms to reproduce at high rates at young ages. It is further proposed that if such organisms are then placed in a protected environment, they will undergo rapid senescence and for that reason will have a short life span, albeit longer than it would be in the wild. The converse is proposed for species that evolved in protected environments in the wild. It is postulated they apportion a considerable amount of energy to somatic maintenance, have a modest rate of reproduction which, however, continues past young adulthood, and have a long life span in the wild, which becomes even longer if placed in a more protected environment than the one in which they evolved. This theory and the database supporting it provide information of great value in the quest for a unifying concept.

"PRIVATE" AND "PUBLIC" MECHANISMS OF AGING

In 1996, George Martin, Steven Austad, and Thomas Johnson published a paper in which they proposed that the proximate mechanisms of aging should be divided into two classes: "private" and "public." The "private" category refers to mechanisms that are idiosyncratic to certain lineages, populations, or species, while the "public" category refers to mechanisms generally operational among a diverse group of species. This classification addresses the apparent paradox that while senescence exhibits general similarities among a broad range of different species, there are many specific differences in the aging phenotype of various species and, indeed, among members of the same species. In my opinion, this classification of proximate aging mechanisms—and the concepts that potentially can be derived from it—are and will be invaluable tools for the conceptual development of biological gerontology.

Martin and his colleagues proposed that mechanisms of senescence associated with the accumulation of mutations with deleterious late-life actions will most likely be in the "private" category. Their proposal is based on the fact that such mutations are neutral in regard to natural selection, neither favoring nor disfavoring evolutionary fitness in young adults (i.e., these mutations are neither selected for nor against). Since genetic drift determines their presence, their occurrence would be expected to vary among lineages. Examples of such mutations in humans, according to Martin and his associates, include the mutation that causes Huntington's disease and the ε4-allele of the apolipoprotein E locus (found in approximately 15% of European adults) which increases the risk of coronary heart disease and late-onset Alzheimer's disease.

The reasoning of Martin and his colleagues is sound and compelling. However, a few additional points should be made. First, this view should be broadened to include all traits that have a detrimental action expressed at advanced ages, rather than being restricted solely to well-defined mutations. For example, essentially all humans show a progressive bone loss with advancing age, which appears to be a species genetic trait (and thus a "private" mechanism) since it does not occur in most mammalian species. Second, it seems likely that antagonistic pleiotropy can also be involved in "private" mechanisms. Indeed, Martin and his colleagues point out that the male sex hormone, so importantly involved in young men generating progeny, has been implicated in the late-life occurrence of prostate cancer and coronary heart disease, a clear example of antagonistic pleiotropy. However, since neither disease is found broadly among mammalian species, this detrimental action should be classified as a "private" mechanism. Finally, although a given "private" mechanism does not by definition occur broadly among species or in some cases among members of a species, it may be quantitatively the most important mechanism underlying senescence in those individuals in whom it does occur, for example, Alzheimer's disease and coronary heart disease in the case of humans.

Martin and his colleagues proposed that "public" mechanisms are most likely to be associated with antagonistic pleiotropic genes. The basis for their view is that there are a relatively small number of processes in which functions are beneficial in early life and harmful in late life, and that such genes will be maintained by strong selective forces and therefore are likely to spread widely among individuals and species. Although this theoretical rationale is reasonable, such genes have yet to be clearly identified, despite much effort to do so. Certainly, if they played a major role in widespread "public" mechanisms, clear evidence for one or more such genes should have been forthcoming by now. Indeed, in my opinion, antagonistic pleiotropy, in so far as it occurs, is likely to primarily underlie "private" mechanisms of senescence.

Furthermore, I believe that "public" mechanisms are likely to be in the domain of The Disposable Soma Theory. All organisms are continuously subjected to damage from intrinsic living processes as well as from a variety of environmental agents, and according to the Disposable Soma Theory, this damage will not be totally repaired. It is my belief that this accumulation of damage occurs in all organisms that undergo senescence, although the rate of accumulation varies among species and among individuals within species. Thus, if common to all species that undergo senescence, the accumulation of unrepaired damage is by definition a "public" mechanism of aging.

UNIFYING CONCEPT

Aging need no longer be considered a biological enigma. The general outlines of the biological processes involved are clear. Why aging occurs is addressed by the evolutionary theory of aging. Although further work is needed to fully verify this concept, it is theoretically sound and, thus far, has been supported by those studies designed to test it. How aging occurs can now be effectively addressed, in general terms, utilizing the concepts of "public" and "private" mechanisms within the context of the evolutionary theory of aging. What remains to be done is to delineate the specific processes involved, and much is already known in this regard.

The underlying causes of aging in the sphere of the "public" mechanisms are the numerous and varied long-term, low-intensity stressors that are inevitably encountered by an organism during its life. By "stressor" is meant an agent or process that tends to cause damage, and "long-term" refers to stressors that occur continuously or frequently during the life of an organism. The origins of these stressors are both intrinsic life processes and environmental agents. However, the extent to which a stressor results in an accumulation of damage depends not only on its intensity and duration, but also on the ability of the protective and repair systems of the organism to counteract the stressor's damaging action. Thus, the rate of aging is determined by the extent to which the damage caused by stressors is not countered by the protective and repair processes.

Do we know the nature of the intrinsic processes that cause low-intensity, long-term stress? Probably many are known, but further work is needed to identify and establish them as stressors causing aging. Oxidative stress is currently a prime candidate. There is strong evidence that oxidative stress results, to a great extent, from intrinsic processes, particularly the organism's use of oxygen in oxidative metabolism to convert energy in fuels to ATP. The reactive oxygen compounds that are yielded by oxidative metabolism can damage a spectrum of vital cellular components, such as chromosomal and mitochondrial DNA as well as proteins and the lipids in membranes. Another long-term intrinsic process that is damaging, and is thus a stressor, involves glycation or glycoxidation resulting from the use of carbohydrate fuel. These intrinsic stressors have been those most studied, but it is likely that other intrinsic processes (as discussed in Chapter 4) are involved.

Many environmental factors can also be long-term, low-intensity stressors that cause accumulation of organismic damage. Examples are infectious agents, thermal insults, and pollutants (natural as well as manmade) in the food, water supply, and air, to name a few. Indeed, the investigation of the role of environmental factors in senescence is in its infancy, though there is already abundant evidence to indicate that environmental factors may often play a more important role than intrinsic factors.

Although long-term, low-intensity intrinsic and extrinsic stressors are continuously present, they need not cause aging if protective and repair processes are able to counteract their damaging effects. Considering oxidative stress as an example, all organisms contain antioxidant compounds (such as vitamin E) and enzymes (such as the superoxide dismutases) that can destroy reactive oxygen compounds before they damage the organism. There are also proteins (such as the heat shock proteins) that protect other proteins, and thus the organism, from oxidative damage. In addition, organisms have many mechanisms for the repair of damage. For example, proteolytic enzymes can remove a damaged protein which, in turn, can be replaced by the biosynthesis of a new undamaged molecule of that protein. In addition, damaged DNA can be repaired by a spectrum of repair processes. The bottom line is that aging occurs when the protective and repair processes do not fully prevent the accumulation of damage caused by long-term, low-intensity stressors. Indeed, the Disposable Soma Theory of Aging postulates that evolutionary forces result in less investment of resources in protective and repair mechanisms than is required for the total prevention of damage accumulation. As a result, the rate of aging may be faster in one species or individual than in another because of greater exposure to long-term, low-intensity stressors, or because of less effective protective or repair processes, or a combination thereof. That senescence does not seem to occur until after sexual maturation is probably due to the fact that evolutionarily adaptive developmental processes promote the biosynthesis of most cellular components, thereby enabling the organism to counter cellular damage as well as further develop its structure and function during that period of life. Moreover, there are periods in young adult life when there is little evidence of damage accumulation; and during such periods, damaging processes surely must be balanced by protective and repair processes. However, sometime after maturation, an imbalance in favor of damage becomes manifest and senescence begins. In organisms such as rats, which evolved in environments that were highly hazardous for them, little energy is apportioned to protective and repair processes and, as a result, the rate of senescence is rapid. In contrast, senescence is slow in humans, who evolved in environments that were far less hazardous for them and thus apportion much more energy to somatic maintenance.

Senescence is not only caused by "public" mechanisms but also by a variety of "private" mechanisms, with particular ones occurring only in some species or some individuals within a species. In a given individual, senesence is due to both "public" and "private" mechanisms and their interactions. Do we know the nature of "private" mechanisms? In general terms, we do. They involve genes with deleterious actions that become manifest only late in life. Although there may be exceptions, the phenotypic effects of most "private" mechanisms are what we recognize as age-associated diseases, which are often unique to a species or group of species or to a subgroup within a species. This does not mean that the stressors involved in "public" mechanisms can not also be involved in "private" mechanisms, but rather that the genetic makeup of the organism results in the stressor producing the unique kind of damage that is characteristic of the particular disease. Atherosclerosis in humans is an example. This condition does not occur widely among species, but it is a major problem for humans in that it underlies age-associated diseases such as coronary artery disease and stroke. There is growing evidence that oxidative stress may well be an important factor in the occurrence of atherosclerosis; specifically, oxidative damage to the low density lipoproteins appears to be a factor in initiating the atherosclerotic lesion. Thus oxidative damage, the same stressor suspected of playing a major role in the "public" mechanisms, is also involved in this "private" mechanism; it is the genetic characteristics of the organism that direct the damage in a specific way, in this case to the formation of atherosclerotic lesions.

The unifying concept is at odds with the classical view of aging in two respects. First, in the classical view, senescence is deemed an entirely intrinsic process; and second, age-associated diseases are not considered a part of "normal" aging. In my opinion, these two classical views hinder our understanding of the biology of aging.

The concept of intrinsicality is flawed for two reasons. First, if senescence results from the long-term accumulation of unrepaired damage, it is immaterial whether the damage is caused by intrinsic processes or by extrinsic (environmental) factors. Second, extrinsic agents cause damage by interacting with biological materials and activities; and in this sense, all damage is intrinsic. Thus one could agree that senescence is entirely intrinsic, but in doing so, the role of environmental factors in intrinsic damage would be downplayed, and that would be unfortunate. Environmental factors are often the most important factors underlying senescence. Moreover, senescence can be modulated by modifying the environment, e.g., by altering lifestyle.

The classical view that age-associated diseases are not an integral part of aging is not conceptually sound if the evolutionary theory of aging is correct. The decrease in the force of natural selection with advancing adult age permits all types of detrimental genetic traits to exist if their damaging action is not manifest until late in life. There is no better example of such traits than age-associated diseases. It is true that some individuals may live to advanced ages without discernible disease, but I believe this is a semantic issue rather than a real difference between this subset of individuals and the rest of the population. Whether to classify an age change as a physiological deterioration or an age-associated disease is arbitrary. The very old individual, free of discernible disease, is nevertheless recognized as frail; arbitrarily, that fraility is referred to as physiological deterioration, and thus, according to the classical view, the person has undergone "normal" aging, i.e., aging in the absence of a disease. The elderly person whose physiological deterioration is recognized by a physician as due to an age-associated disease and the elderly person whose physiological deterioration is recognized only as old age are both, in my view, experiencing the effects of senescence.

The unifying concept provides an overall understanding of the elements of the biology of aging. Of course, further research will be needed to obtain a detailed knowledge of the specific processes underlying senescence in a particular species. The unifying concept will be invaluable in the design of such research and in the utilization of currently available information for this purpose. As a case in point, the available information provides considerable insight into the basic biological processes in human aging when interpreted within the framework of the unifying concept. For example, the question of the relative importance of intrinsic versus extrinsic processes in human aging can be addressed. Rowe and Kahn have presented evidence that senescent deterioration in humans is primarily a function of lifestyle. Thus intrinsic stressors *per se* do not appear to be a major factor in human aging, but the interaction of lifestyle (extrinsic factors) with intrinsic factors and/or extrinsic factors *per se* appear(s) to be the dominant factor or factors. Also, diseases such as coronary heart disease, stroke, and Alzheimer's disease are major causes of human senescent deterioration, which means that "private" mechanisms play a dominant role in human aging; however, "public" mechanisms, such as oxidative stress, may well be involved in these "private" mechanisms. Clearly, while much more remains to be learned about human aging, a basic understanding is in hand. Moreover, each species must be assessed separately because it is clear from the available information that the aging of each species is, in some way, unique to that species. Indeed, the same can be said for each individual within a species. In conclusion, the basic biological nature of aging can no longer be considered an enigma, though, of course, a detailed understanding of senescent processes is still lacking, even in such species as rats, mice, and humans which have been the major focus of biogerontologic research.

ADDITIONAL READING

Holliday, R. (1995). *Understanding aging.* Cambridge, England: Cambridge University Press.

Kirkwood, T. B. L. (1990). The disposable soma theory of aging. In D. E. Harrison (Ed.), *Genetic effects on aging II* (pp. 9–19). Caldwell, NJ: Telford.

Martin, G. M., Austad, S. N., & Johnson, T. E. (1996). Genetic analysis of ageing: Role of oxidative damage and environmental stresses. *Nature Genetics, 13,* 25–34.

Masoro, E. J. (1996). The biological mechanism of aging: Is it still an enigma? *Age, 19,* 141–145.

Miller, R. A. (1997). When will the biology of aging become useful? Future landmarks in biomedical gerontology. *Journal of the American Geriatrics Society, 45,* 1258–1267.

Rose, M. R., Vu, L. N., Park, S. U., & Graves, J. L., Jr. (1992). Selection on stress resistance increases longevity in *Drosophila melanogaster. Experimental Gerontology, 27,* 241–250.

6

The Human Aging Phenotype

When first meeting a person, most of us automatically judge his or her age. This judgment is based on observable characteristics, and the complex of those characteristics associated with age is called the human aging phenotype. These include facial appearance, hair pigment and distribution, posture, fat distribution, pattern of walking, and a host of other clues. Of course, our judgment is in no way exact, because of the wide individual variation in each of these characteristics with advancing age. Nevertheless, on average, we are reasonably good at assessing the approximate chronological age of those we meet.

In addition to characteristics perceptible to the naked eye, the human aging phenotype includes characteristics discernible with tools for measuring physical and chemical parameters, such as blood pressure, blood sugar level, and body fat content. This chapter will focus on the age changes in anatomical and physiological characteristics and on the occurrence of age-associated diseases. Although some of these changes may be a cause of senescence, most are probably the result of it, though the relationship of any phenotypic characteristic to aging is far from clear. Indeed, some phenotypic changes may have little to do with organismic senescence. For example, there is no evidence that graying of the hair increases a person's vulnerability or decreases the ability to function. Thus, although associated with advancing calendar age, graying of hair probably does not belong in the category of senescence. On the other hand, gray hair is caused by an age-associated loss in function of a specific type of cell, namely a loss in the ability of melanocytes to provide the hair follicle with pigment. Thus, in this case, as in many others, the relationship of phenotypic change to senescence has ambiguities which, given our current state of knowledge, cannot be fully addressed.

BODY STRUCTURE AND COMPOSITION

With advancing adult age, there are general changes in human body structure and composition. Describing these changes is a good starting point for this discussion of the human aging phenotype.

Height

By age 70, the height of men and women is some 2.5% to 5% below its peak level. This decrease begins at about age 25 in men and age 20 in women. In a longitudinal study in which people in the age range of 55 to 64 were followed for the next 11 years, the average loss in height was one-half inch for men and one inch for women. The loss of height is due primarily to the compression of the cartilaginous discs between the vertebrae and to a loss of vertebral bone.

Weight

Body mass, commonly referred to as body weight, increases in most American men from age 20 until middle age, followed by a decline at advanced ages, particularly after age 70. For most American women, weight increases from age 20 to 45, after which it remains stable until about age 70 and then declines with increasing age. It is useful to further assess age changes in weight by a two-component model consisting of the lean body mass and the fat mass.

Lean Body Mass

The difference between the total body mass and the mass of adipose tissue, commonly referred to as the fat mass, is called the lean body mass. During adult life, lean body mass, which is greater in men than in women, declines by about 0.3% per year in men and 0.2% per year in women. Much of this decrease is due to the loss of muscle mass, but a loss of bone and other structures is also involved. It has been claimed that the loss of lean body mass is due to a more sedentary lifestyle. It is true that people who continue to engage in athletics until advanced ages have a greater lean body mass at any given age than those of similar size who are sedentary. However, even in athletes, there is a progressive loss in lean body mass with advancing adult age.

Fat Mass

Most of the body fat resides in the adipose tissue, which is primarily located beneath the skin (subcutaneously) and around organs, particularly the abdominal organs (referred to as abdominal visceral fat). Adipose tissue is composed mainly of adipocytes, which are large cells that have a central droplet of fat surrounded by a thin rim of cytoplasm. The size of the adipocyte changes as fat is added to or removed from the central droplet.

The percentage of fat content of the body increases with increasing adult age, and the extent of increase varies among individuals within a population and among different populations. In the United States, it has been found to be, on average, about 18% in 30-year-old men and 26% in 70-year-old men, and about 24% in 24-year-old women and 36% in 70-year-old women. (These values may be changing because of what could be termed the current epidemic of obesity in the United States.) With increasing age, there is also a change in distribution of adipose tissue, with fat tending to accumulate in the abdominal region, primarily around the viscera, in both sexes, though it is greater for men than for women. The accumulation of visceral fat has been implicated as a risk factor for age-associated cardiovascular disease.

An increasingly sedentary lifestyle with advancing age is a factor responsible for increased body fat content. Those who continue as athletes into advanced ages have a smaller increase in body fat content, but even in these people, some increase does occur. It is important to note that physical activity retards the age-associated redistribution of body fat to the abdominal visceral region.

Cellularity

As discussed in prior chapters, the human body has trillions of cells, which include many different types with highly specialized functions.

Cell Atrophy and Loss

Water is the major component of most cells, and total intracellular water decreases with increasing adult age in both men and women. This finding suggests either a decrease in the number of cells or in the size of cells (i.e., cell atrophy) or both. The answer is both, but it is important to note that particular cell types at specific anatomical sites are involved, rather than any generalized loss or atrophy of cells. For example, both a loss of cells and cell atrophy have been noted in skeletal muscle and brain, but not in all muscles, nor in all brain regions. Thus assessment of the possible role of cell loss or atrophy in age-associated functional impairments must focus on particular cell types in specific regions of organ systems.

Hypertrophy and Hyperplasia

At some anatomical sites, there is not a decrease but rather an increase in the size of the cells (hypertrophy) or in the number of cells (hyperplasia) with advancing adult age. Examples are hypertrophy of heart muscle cells and hyperplasia of the cells of the prostate. The increase in the size or number of cells may not be beneficial but, rather, may cause functional problems. Such is the case in the two examples cited.

Neoplasia

An unregulated (or at best, poorly regulated) and progressive increase in the number of cells, usually different in character from normal cells, is referred to as neoplasia. Benign tumors and malignant ones (cancers) are the result of this process. Although cancers are not restricted to the elderly, their incidence increases with increasing age, the prevalence among people over 50 being much greater than among those younger than 50.

Extracellular Matrix

The cells of the body are embedded in extracellular matrix which serves as a scaffolding for the tissues and organs of the body. In addition to playing a key role in the structure of organs, extracellular matrix also interacts with the cells embedded in it, so as to influence their functional characteristics. The changes that occur with age in the macromolecules of the extracellular matrix were discussed in Chapter 4. These changes affect the structural characteristics of the tissues and organs; they underlie the loss in tensile strength, as well as the decrease in compliance (distensibility) and the decrease in elasticity (resiliency) of specific organs and organ systems, a problem that will be considered in this chapter during the discussion of specific organ systems.

SKIN

Because skin is easily observable, it is used by all of us to assess age. However, before discussing age changes that occur in the skin, it is necessary to briefly consider its structure.

The skin is composed of three layers: epidermis (the outermost), dermis (the middle layer), and subcutaneous fat (the innermost). Figure 6.1 is a schematic presentation of the microscopic structure of the skin of a young adult.

More than 90% of the cells in the epidermis are keratinocytes. These are continuously formed by the mitotic activity of cells near the basement membrane that separates the epidermis from the dermis. The newly formed keratinocytes migrate towards the skin's surface as they gradually differentiate to ultimately become flat cells that are tightly packed and adhere to each other. There are about 10 to 15 layers of these flat keratinocytes at the skin surface where they serve as a barrier, impeding loss of water from the body and entry of chemicals and microbial pathogens into the body. The keratinocytes on the skin surface are constantly being shed and replaced by keratinocytes that have newly migrated to the skin surface.

In addition to keratinocytes, some 2% to 4% of cells in the epidermis are melanocytes, which produce pigment that is transferred to the keratinocytes. The remaining 1% to 2% of the cells in the epidermis are Langerhans cells, which have immunological functions.

Below the basement membrane lies the dermis, which is comprised of supportive connective tissue, microvascular structures (very small blood vessels) of both the superficial and deep vascular

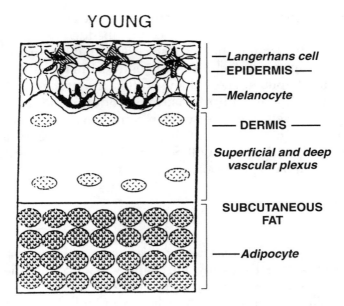

Figure 6.1 Schematic drawing of the skin of a young person. The three layers of the skin are depicted: epidermis; dermis; and subcutaneous fat. The wavy black line separating the epidermis from the dermis denotes the basement membrane. Cells of the epidermis: the open oval and round structures and the flat structures near the skin surface denote keratinocytes at various stages of differentiation; the black structures near the basement membrane with projections into the epidermis denote melanocytes; the star-like structures denote Langerhans cells. The ovals in the dermis denote vascular plexuses. The circles in the subcutaneous fat denote adipocytes.

From "Physiology, Biochemistry and Molecular Biology of the Skin," (2nd. ed., p. 1426), L. Goldsmith (Ed.), 1991. New York: Oxford University Press. Copyright 1991 by Oxford University Press. Adapted with permission.

plexuses, nerve fibers, and appendageal structures (hair follicles, sweat glands, and sebaceous glands). Below the dermis is the subcutaneous fat, whose major cell type is the adipocyte.

Intrinsic Aging

In nonsmokers, there are age changes that occur in the skin areas protected from exposure to the sun, and this is referred to as intrinsic aging of the skin. To some extent, this is probably a misnomer:

Although sun exposure and smoking are the major extrinsic factors influencing skin aging, they may not be the only ones.

The thickness of the barrier of flat keratinocytes at the surface of the epidermis does not change with age, but the rate of shedding these cells decreases, as does their rate of replacement. Thus, in the elderly, these cells remain longer at the surface of the skin, which increases the likelihood that they will accumulate damage. The number of melanocytes decreases with increasing adult age, with a 10% to 20% reduction each decade; thus, the skin of the elderly, if exposed to sunlight, is less protected from the damaging action of ultraviolet light. At advanced ages, the epidermis contains 20% to 50% fewer Langerhans cells than at young ages; this may increase susceptibility to skin cancers. With increasing age, the structure of the basement membrane alters, decreasing the extent of interaction between the dermis and epidermis, and thus increasing the likelihood of injuries causing the two layers to separate.

The thickness of the dermis is about 20% less in the elderly, and the dermis is stiffer and less malleable, making it more vulnerable to injury. Many of these changes in the dermis are due to alterations in the fibrous proteins (collagen and elastin) of its extracellular matrix. There is an age-associated decrease in collagen content as well as changes in its structure. The fine wrinkling of the skin of the elderly probably results from alterations in the structure of elastin. There is also a decrease with age in the small blood vessels of the dermis, as well as a reduction in the number of hair follicles. In addition, there is an age-associated reduction in both the number and functional capacity of the sweat glands. Also, the ability of the sebaceous glands to secrete sebum (a greasy lubricating substance) is decreased.

Subcutaneous fat (the innermost layer of the skin) increases with advancing age in some regions of the body, such as the waist in men and the thighs in women; it decreases in other regions, such as the face, hands, shins and feet. The loss of facial subcutaneous fat plays a big part in facial changes that characterize the elderly.

Extrinsic Aging

Areas of skin exposed to sunlight show more deterioration with advancing age than do those that are protected, a phenomenon called photoaging. Smoking also accelerates the age-associated deterioration of the skin. Although other extrinsic factors undoubtedly influence skin aging, photoaging and smoking appear to be, by far, the major ones.

Photoaging

Chronic photodamage (sun damage) is estimated to cause more than 90% of age-associated skin cosmetic problems. It leads to coarseness of the skin, dilation of groups of small cutaneous blood vessels, irregular pigmentation, and deep wrinkles. It also causes a decrease in the number of epidermal Langerhans cells. Damage to elastin and collagen in the dermal extracellular matrix underlies many of the cosmetic problems.

Smoking

Aspects of skin aging are accelerated by smoking. In particular, smoking has been found to promote wrinkling of the facial skin. Moreover, there is a synergistic action between smoking and sun exposure, resulting in a marked acceleration of wrinkling.

Age Changes in Skin Physiology

In addition to the well-documented changes in the structure of skin, are there changes with age in the physiological functions of skin? The barrier function of the skin has been the subject of much study, and it appears that the loss of body water through the epidermis is decreased with increasing age. Interestingly, there also appears to be a decrease in the ability of substances to enter the body through the skin; the extent of this decrease relates to the molecular structure of the entering substance.

The immune function of the skin is also altered with advancing age, as might be expected from the decrease in the number of Langerhans cells. Delayed hypersensitivity reactions to common antigens, such as those of candida (a genus of yeast that causes eczema-like lesions) and mumps, are less vigorous in the skin of the elderly. There is also a decrease with advancing age in immediate hypersensitivity reactions (i.e., what most people recognize as classic skin allergies) in areas not exposed to the sun. However, in sun-exposed areas, allergic reactions may be increased, because photoaging increases the number of mast cells in the dermis; these cells secrete histamine in response to allergens, thereby causing many of the disagreeable characteristics of skin allergies.

The inflammatory response of the skin to harsh chemicals, such as kerosene, is less intense in the elderly. Because inflammation alerts the individual to noxious substances, the attenuation of this response makes older people more susceptible to harm from such substances.

The healing of skin wounds involves the following sequential events: inflammation, cellular proliferation, and formation and maturation of extracellular matrix. With increasing adult age, all stages of wound healing are impaired.

Skin Cancer

After age 30, there is an exponential increase in the incidence of skin cancer with increasing age. About 90% of these skin cancers occur in the approximately 10% of the skin that is habitually exposed to the sun, which indicates that aging alone does not appreciably predispose skin to cancer in the absence of an extrinsic factor such as sunlight.

Pressure Sores

Although not as common as skin cancers, pressure sores (also referred to as decubitus ulcers or bedsores) are a major concern of geriatricians. Pressure sores can occur in people of all ages, but their prevalence increases so markedly with increasing age that people over 70 account for 70% of those with this lesion. In its mildest form, there is a redness of a skin area which, if the lesion becomes more severe, progresses to a loss of epidermal and dermal structures; ultimately there is full-thickness skin loss, and tissue necrosis occurs in the most severe lesions. The main causal factor is prolonged pressure on a particular area because of reduced mobility. Additional factors are friction and moisture, particularly moisture due to urinary or fecal incontinence. Obviously, there is a greater incidence of pressure sores in the elderly, because they are more likely to have medical problems requiring prolonged periods in bed and thus sustained pressure on particular skin areas and, in many cases, exposure to prolonged periods of moisture, due to the increased prevalence of urinary and fecal incontinence. In addition, the elderly are predisposed to pressure sores because of the age changes in their skin, which have been described above.

MUSCULOSKELETAL SYSTEM

The musculoskeletal system is comprised of the skeletal muscles, bones, and joints. This system plays a key role in movement, posture, and the ability to do physical work. These functions deteriorate at advanced ages; in some individuals, the extent of this deterioration is so great that it reduces the quality of life.

Skeletal Muscle

Skeletal muscles are those muscles that are under voluntary control and attached to the skeleton. There are two other categories of muscle: Cardiac muscle, an important component of the heart; and smooth muscle, a component of blood vessels and certain organs and organ systems. Discussion of these other categories of muscle will await the presentation of the relevant organs and organ systems.

The skeletal muscles are composed of a large number of muscle cells (called muscle fibers) bound together by connective tissue and connected by fibrous tendons to bones or other structures. The number of fibers varies among different muscles, ranging from less than 100,000 to 1,000,000 or more. The lengths of muscle fibers also vary, fibers in some muscles being as long as 30 cm (12 inches). Skeletal muscle fibers have the organelles and proteins found in most cells, but, in addition, contain membranous structures and proteins needed for the cell to contract (i.e., to shorten and/or develop force) and to relax. The proteins, actin and myosin, are organized in filaments that are arranged so that when the two proteins interact, the muscle fiber contracts. The interaction of actin and myosin is regulated by a complex system that involves other proteins, such as tropomyosin and troponin, and a membranous cellular organelle called the sarcoplasmic reticulum, which regulates the concentration of calcium ion surrounding the regulatory and contractile proteins. It is this concentration of calcium ion that determines whether actin and myosin interact to cause contraction of the fiber; i.e., high calcium ion concentration results in contraction, and low concentration in relaxation. Of course, muscle contraction and relaxation require the expenditure of energy, which comes from the ATP generated in the muscle fiber by the metabolism of fuels.

Age Changes in Muscle Structure and Function

A decrease in skeletal muscle mass (referred to as sarcopenia) occurs with increasing adult age. It is generally agreed that there is an age-associated decrease in the number of muscle fibers in many, but not all, skeletal muscles. Associated with this decline in muscle mass is a decrease in muscle strength (the maximum force that can be produced by contraction of the muscle). Between 30 and 80 years of age, muscle strength declines by about 30% for the arm muscles and about 40% for the back and leg muscles. With the decrease in muscle force, there is a decrease in muscle power output (the product of muscle force X speed of movement); as a result, there can be impaired ability to perform everyday tasks, such as rising from a chair or climbing stairs.

Age and Muscle Usage

With increasing adult calendar age, most people adopt a more sedentary lifestyle. Skeletal muscle is very plastic (modifiable), and disuse will lead to a decrease in muscle fiber size as well as a change in functional characteristics. An unresolved question is the extent to which disuse is responsible for the age-associated loss of skeletal muscle fibers.

It has been found that 70-year-old men who have been physically active or have undergone strength training can develop a muscle force as great as sedentary young adults. This stems from developing an increased muscle fiber size to compensate for the reduced number of fibers. Thus, some of the age changes in skeletal muscle function with advancing age are clearly due to a lack of usage. The effectiveness of exercise programs to increase muscle mass and strength in the elderly will be discussed further in Chapter 7.

Neuromuscular Function

For activities such as walking, writing, and using tools (and indeed, all functions using skeletal muscles), the contraction of skeletal muscle fibers must be controlled by the central nervous system (brain and spinal cord). The central nervous system is connected to the skeletal muscle fibers by motor axons, nerve fiber projections of motor neurons (nerve cells) in the spinal cord and brain stem. Each motor axon branches out in the vicinity of the muscle fibers, enabling it to contact several muscle fibers; the motor axon and the muscle fibers it supplies are called a motor unit. The contact between a terminal branch of the motor axon and a skeletal muscle fiber is called the neuromuscular junction; there is a very small fluid-filled space between the cell membrane of the axon and that of the muscle fiber.

The central nervous system signals muscle fibers to contract by sending nerve impulses along the motor axons that supply the muscle fibers. A nerve impulse is an electrical disturbance, called an action potential, that is propagated along the length of the axon. The action potential causes the axon to release acetylcholine (a neurotransmitter) into the fluid-filled space of the neuromuscular junction. The acetylcholine reacts with receptors on the membrane of the muscle fiber, and this interaction results in the generation of an action potential that travels the length of the muscle fiber. This action potential causes the muscle fiber to contract.

With increasing adult age, there is a loss of motor neurons. As a result, some muscle fibers are no longer supplied by a motor axon and are said to be denervated. However, most, but not all, of the denervated muscle fibers subsequently receive axon branches from the remaining neurons and are said to be reinnervated. It seems likely that, at least in part, the loss of skeletal muscle cells with increasing age is secondary to denervation. Moreover, the reinnervation of muscle fibers by the remaining motor neurons means that a motor axon is supplying a larger number of muscle fibers, a phenomenon referred to as an increase in the size of the motor unit. Larger motor units reduce the ability to carry out fine movements.

Bone

The bones of the body collectively form the skeleton, which supports the body, enables locomotion, and protects the vital organs. About 65% of the dry weight of bone is mineral, primarily calcium phosphate salts; about 35% is extracellular matrix, 90% of which is collagen. However, there are also three principal cell types in bone: the osteoblasts, which are involved in bone formation; the osteoclasts, which are involved in bone resorption; and the osteocytes, the function of which is unknown.

Bone continuously undergoes resorption and reformation, a process known as bone remodeling. Remodeling begins by osteoclasts attacking a site on bone to form a small cavity in the bone, which is then lined by osteoblasts that form bone to refill the cavity. In the young adult, the process of bone resorption is balanced by bone reformation, so that remodeling causes no change in the amount of bone.

Aging and Bone Loss

With advancing adult age, the balance during remodeling shifts in favor of bone resorption. Thus, in the case of every population that has been studied, an age-associated loss of bone mass has been found to occur. Nearly all bones in the skeleton are so affected, though to varying degrees. Bone loss is greater in women than in men, and the rate of loss accelerates in women after menopause; such acceleration has been found to be due to the low postmenopausal levels of estrogen (female sex hormone). This rate of bone loss is most rapid during the 10 years following cessation of menses and slows after that. In women, bone loss in the vertebrae begins as early as the third decade of life, but bone loss in the legs and arms does not occur until the sixth decade. Men start to lose bone at later ages than do women, and in some men, the amount of loss is trivial. Many theories have been proposed in regard to the basic mechanisms responsible for age-associated bone loss, but as of now there is no strong evidence in support of any of them.

Osteoporosis

Low bone mass and increased susceptibility to bone fractures from minor trauma is a medical condition called osteoporosis. Some 15 to 20 million Americans age 45 and older are afflicted with this condition, and it causes about 1.3 million bone fractures annually in the United States. There are two major types of age-associated osteoporosis: the postmenopausal type I, and the senile type II.

Type I is more prevalent, relates to the menopause, occurs six times more frequently in women than in men, and is frequently associated with fractures of the vertebrae and wrists. It is due to an increase in the rate of remodeling of bone, with the increase in resorption somewhat greater than the increase in bone formation. The lower incidence in men is due to three factors: a greater bone density upon reaching maturity; the shorter life expectancy of men; and the fact that men do not have a rapid endocrine change equivalent to menopause.

Type II osteoporosis occurs primarily in people over 70, affecting about twice as many women as men. It results mostly in vertebral wedge fractures and hip fractures, and appears to be due to an age-associated decrease in bone formation.

Risk Factors for Osteoporosis

While age-associated bone loss is inevitable, the skeletal mass at maturity is a predictor of those most likely to suffer a fracture due to osteoporosis later in life. As already noted, the larger skeletal mass of men is one reason for the gender difference in osteoporotic fractures. Black women also have a larger skeletal mass than do Caucasian and Asian women, which is the major reason that they suffer fewer osteoporotic fractures. High calcium intake and/or physical exercise while young enhance(s) the bone mass attained at maturity. Even during adult life, exercise is likely to be a factor in maintaining bone mass, as indicated by the loss of bone mass during extended periods of bed rest or during weightlessness in space. Adequate dietary calcium is also necessary for adults to minimize age-associated bone loss, and it appears that the calcium intake of many adult Americans may not be adequate in this regard. Epidemiological studies indicate that alcoholic beverages, smoking, and caffeine consumption are associated with an increase in the age-associated loss of bone and, therefore, are risk factors for fractures due to osteoporosis.

Joints

Joints are the connections between the bones. There are several types of joints but those of importance to gerontology and geriatrics are in the class called synovial joints. The structures of such joints are illustrated in Figure 6.2. This type of joint permits relatively free movement of the bones joined by the joint. The ends of the bones forming the joint are covered with cartilage (a smooth, rubber-like covering that reduces friction and absorbs shock during movement), and the bones are connected by a joint capsule and ligaments. The capsule is composed of a fibrous layer, the inner surface of which is

The Human Aging Phenotype

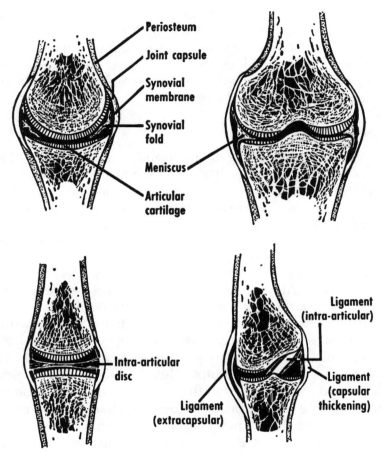

Figure 6.2 Synovial joints. The size of the joint cavity is exaggerated in order for it to be clearly visible.

From "Basic Human Anatomy: A Regional Study of Human Structure," by R. O'Rahilly, 1983, p. 12. Reprinted with permission of the author.

lined with a vascular connective tissue called the synovial membrane. This membrane produces the synovial fluid that fills the joint cavity.

Joint flexibility is defined as the extent to which bones linked at the joint are able to move before being stopped by bony structures, tight ligaments, tendons, or muscles. Loss of flexibility reduces the movements that can be made by a joint; and it also predisposes the joint, and/or the muscles crossing the joint, to injury, including muscle strain and tendon and ligament damage. With increasing adult age, there is a loss of joint flexibility, thus reducing the range of motion and increasing the possibility of damage to joints and the muscles crossing the joint. Carefully designed exercise programs can improve the mobility of joints and thus, to some extent, counter age-associated loss of joint flexibility. However, it must be emphasized that inappropriate exercise regimes can be harmful.

Osteoarthritis

Arthritis refers to joint inflammation, and osteoarthritis is a degenerative disease of the joints. It affects, to some degree, about 80% of people 65 or older; women are more seriously affected than men. In this disease, the cartilage in the joint changes in consistency, cracks, and wears away, ultimately exposing the bone surface to other bones. With time, further changes in bone may occur, such as the development of bone spurs, abnormal thicknesses, and fluid-filled pockets. Periodic or chronic inflammation can occur, which is often accompanied by pain. Although clearly an age-associated problem, the causes of osteoarthritis are not fully understood. In many cases it appears that physical damage—due to accidents or overuse during work or sports—that occurred earlier in life may be involved. While osteoarthritis can occur in any joint, most commonly it affects the joints of fingers, knees, and hips. This disease is one of the major causes of reduced mobility in the elderly. Exercise programs are helpful for those suffering osteoarthritis, but they must be of a type that does not further damage the joints. In this regard, exercise programs utilizing heated swimming pools have been found to be very helpful.

Rheumatoid Arthritis

This type of arthritis is an autoimmune disease. It involves inflammation of the joints and begins with symmetrical pain and swelling of the joints in the hands and feet. Most frequently it occurs in midlife, though onset in late life is not rare. Most of those suffering from rheumatoid arthritis will have intermittent periods of active disease and periods of partial or complete remission. The prevalence of rheumatoid arthritis increases with age in both men and women. At ages below 60, it affects mostly women but at more advanced ages, men are increasingly afflicted. In its most severe form, this disease leads to difficulties in walking and the activities of daily living.

Gout

This type of arthritis involves the formation of uric acid crystals in the synovial fluid. These crystals are believed to play a key role in the inflammation in the joints, which is the major characteristic of this disease. The joint is hot, red, swollen and extremely painful. Although the joint in the big toe is the one most commonly involved, any joint may be affected. Gout is more common in men than in women, whose age of onset is older than men; peak age of incidence in men is the fifth decade of life. Exacerbations and remissions characterize the course of the disease. The major causal factor of gout is the plasma uric acid concentration, and it is the elevation of its concentration with increasing age that probably accounts for the age-associated characteristic of this disease.

NERVOUS SYSTEM

The nervous system is composed of the central nervous system (i.e., the brain and spinal cord) and the peripheral nervous system (i.e., the nerves that run to and from the central nervous system). Figure 6.3 schematically depicts the gross structure of the brain. The forebrain is dominated by cerebral cortices which overlie subcortical structures such as the thalamus, hypothalamus, and corpus striatum. The brain stem, which includes the midbrain, pons, and medulla, connects the forebrain to the spinal cord. The spinal cord (not labeled in Figure 6.3) is continuous with the brain stem, starting just below the medulla and continuing almost to the tailbone. It is clear from the figure that the brain contains many different structures; while each has unique functions, this discussion will be limited to those related to aging.

The neuron (nerve cell) is the fundamental unit of nervous system function. About 100 billion neurons make up the human nervous system. They vary so greatly in structure that there are an estimated 10,000 different types. Examples of the varieties of structures are shown in Figure 6.4. Despite the variation in structure, every neuron has a cell body and projections called dendrites and axons. (Most of these projections are short, about 1 mm or less, but a few can be quite long; e.g., some axons run from the spinal cord to the muscles in the feet.)

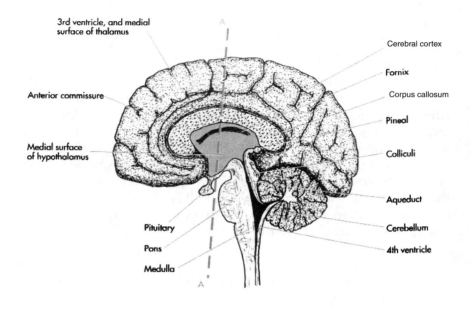

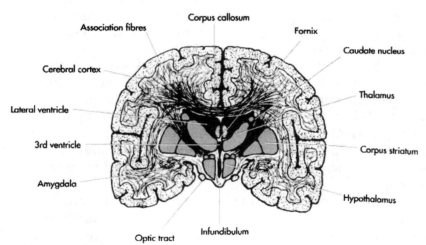

Figure 6.3 Semi-schematic drawings of the human brain. Upper drawing is a view of the brain that is referred to as a sagittal section; i.e., cutting the brain along a mid-plane from top of the head to the beginning of the neck thereby showing the medial aspect of the brain. Lower drawing is transverse section through the plane A-A shown in the upper drawing.

From "Neurophysiology," (3rd ed., p. 8), by R. H. S. Carpenter, 1996. London: Arnold. Copyright 1996 by Arnold. Reprinted with permission.

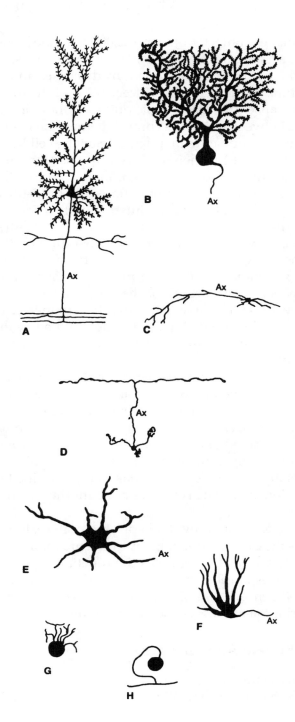

Figure 6.4 Schematic representation of some typical neurons. Symbols denote the following: ax, axon; A, pyramidal neuron; B, flask shaped Purkinje neuron; C, stellate neuron; D, granular neuron; E, multipolar anterior horn neuron; F, multipolar sympathetic ganglion neuron; G, multipolar parasympathetic neuron; H, pseudounipolar dorsal root ganglion neuron.

From "Functional Neuroanatomy," (p. 5), by A. K. Afifi & R. A. Bergman, 1998. New York: McGraw-Hill Companies. Copyright 1998 by The McGraw-Hill Companies, Inc. Reprinted with permission.

A typical neuron makes contact with about 2,000 other neurons, and the points of contact are called synapses. The synapses are similar in structure to the neuromuscular junctions discussed in the Musculoskeletal section. A brief simplified description of the functioning of a single synapse provides a basic outline of how information is transmitted between neurons. A branch of the axon of one neuron (neuron A) makes contact with a dendrite or the cell body of another neuron (neuron B). When a nerve impulse traveling along the axon reaches the end of the axon branch, it causes the axon to release a neurotransmitter into the minuscule fluid-filled space of the synapse. The neurotransmitter interacts with a specific receptor of the dendrite or cell body of neuron B and it has either an excitatory or inhibitory effect on neuron B, depending on the chemical nature of the neurotransmitter. (An excitatory effect means that it promotes the generation of nerve impulses by that neuron, while an inhibitory effect means it restrains the generation of nerve impulses by that neuron.) Of course, in reality, the frequency of nerve impulse generation by neuron B would be determined by the activity of the thousand or so synapses (excitatory and inhibitory) at a given moment in time.

In addition to the neurons, the nervous system contains about 500 billion glial cells. There are several types of glial cells, and it is likely that all their functions have yet to be uncovered. It is known that some make myelin which covers the surface of axons, thereby influencing the rate at which axons conduct nerve impulses. Other functions of glia include: regulating the composition of brain extracellular fluid; influencing metabolic activities of neurons; and protecting the nervous system from intruders, such as proteins not normally present in the central nervous system.

The nervous system keeps an individual apprised of the external environment, enables the carrying out of many different motor functions (e.g., walking, sewing), and regulates the functioning of the organ systems. It is also responsible for the so-called higher functions such as memory, intelligence, learning, and creativity. The influence of aging on these various activities will be discussed presently.

General Nervous System Age Changes

Over the years, many claims have been made about generalized age changes in the nervous system. When carefully studied with

Structural Changes

Brain weight has been reported to decline with advancing adult age; this conclusion was based on measurements made after death. However, there is concern about the validity of this claim: Not only were the individuals studied of different ages but they were from different cohorts; i.e., the oldest were born around 1900 and the youngest 50 years or so later. Studies in which individuals of different ages are also from different cohorts are called cross-sectional studies, and these have many pitfalls. For example, it is known that the brain weight of young people who died in 1870 was less than that of the same age who died in 1970. Thus clearly there is a cohort effect, which points up the difficulty of determining how much of the difference reported for brain weight of the old and the young is due to age and how much to a cohort difference. This interpretational problem would be eliminated if brain weight could be measured at different times during the life of an individual. Of course, it is not possible to weigh the brain of a living person. However, modern imaging technology enables the measurement of the volume of the brain (a good surrogate for weight) in the living individual. This would make it possible in the future to measure brain volume repeatedly during the life of an individual, and thus ascertain if, indeed, there are age changes in brain weight.

The central nervous system is cushioned by the cerebrospinal fluid; most of it is in the ventricles (large chambers) that are surrounded by the brain. Based on cross-sectional studies using imaging technology, it appears that the brain makes up about 90% of the cranial cavity until about age 50; after that, with advancing age, it occupies a progressively smaller percentage of the cranial cavity. An increase in volume of the cerebrospinal fluid in the ventricles is about the same magnitude as the reduction in brain volume.

In the 1950s, it was believed that large numbers of neurons were lost with increasing adult age. This belief has not been supported by modern studies. Now most experts agree that there is little generalized age-associated loss in neurons, although some localized brain regions do suffer substantial loss. Rather than generalized neuron loss, there is neuron atrophy; as a result, the number of large neurons has been found to decrease and the number of small neurons to increase with increasing age. The number of several types of glial cells increases with adult age; however, in one type, there is an increase in cell size rather than an increase in number of cells.

In many regions of the brain, the number and complexity of dendritic projections from neurons decrease with increasing age, and this results in a decrease in the number of synapses. This loss is such that people of age 75 have about 20% fewer synapses than those of age 40. While it is felt that synaptic loss probably accounts for many of the functional deficits associated with aging, this has not been proven.

Blood Flow and Metabolism

The brain generates the adenosine triphosphate (ATP) required for its energy needs by utilizing oxygen to metabolize glucose fuel. Thus the brain needs a continuous supply of oxygen and glucose, which are brought to it by the blood. There appears to be little change in blood flow to the brain or in the metabolic use of oxygen and glucose by the brain until the eighth decade of life; and even at 80-plus, the decreases noted in some individuals may relate to disease processes. However, these data do not take into account the complexity of brain function. Specific cells at particular times, for example, require the increased use of glucose. Thus measuring total blood flow and the use of oxygen and glucose may not reflect a reduced ability of particular cells to meet their immediate energy needs. Fortunately, the technology is now at hand to answer this kind of question, though much time-consuming work will be required to fully address this important issue.

If they occur at all, changes in blood flow to the brain are subtle in those elderly not suffering from a brain disease. However, the changes are dramatic in those suffering from a stroke. Stroke refers to a sudden or relatively rapid occurrence of inadequate blood flow to the brain, resulting in disturbed brain function; it is caused by the blockage or rupture of a brain blood vessel. Although a stroke may occur at any age, it is most common in the elderly; from middle age onward, there is a doubling in the frequency of stroke in each successive decade of life. In the United States, the incidence of strokes is about 500,000 to 700,000 per year, and more than 150,000 deaths are annually attributed to strokes. More than half the survivors have functional impairments, ranging from the inability to function in the work force to loss of ability to carry out activities of daily living (ADL). Stroke can also lead to dementia, and it is estimated to account for 10% to 20% of all dementia cases, making it second to Alzheimer's disease as a cause of dementia.

Sensory Functions

The sensory systems provide information about the external and internal environment by sending nerve impulses from sense organs or receptors, via sensory nerve fibers, to the central nervous system. The sense organs can be as simple as the ending of a nerve fiber, or as complex as the ear and the eye. (It should be noted that physiologists use the word "receptor" in two rather different ways: 1. the cells in a sense organ that respond to the external or internal environment; 2. the molecular structures of a cell that interact with a neurotransmitter or hormone or cytokine.)

Skin Sense

The skin has receptors that respond to stimuli that give rise to the sensations of touch, pressure, and vibration. Several specific types of receptors appear to be involved, including relatively complex structures (Pacinian corpuscles, Meissner corpuscles, and Merkel disks) as well as free nerve endings. Although the skin also has receptors that respond to temperature, there is much uncertainty about their structure. Sensations of pain can arise from heat, cold, mechanical damage and chemical irritants; it appears that free nerve endings are the receptors involved. The information from sense organs is sent to the central nervous system where it can result in a response not involving higher centers (e.g., withdrawal of the hand from a hot stove involves primarily the spinal cord) although the information also goes to higher centers, including the cerebral cortex, where sensations are registered.

The number of Pacinian corpuscles and Meissner corpuscles decrease throughout life, so that by late life there are far fewer than in the young adult. Little change occurs with age in the number of Merkel disks and free nerve endings. There is a decrease in sensitivity to touch (particularly in the hand region) and in the ability to distinguish between two spatially distinct points of contact. High-frequency vibration is sensed less well with increasing age by the Pacinian corpuscles, particularly in the feet and legs. Although the ability to detect the onset of pain is not affected by age, there is debate as to whether the elderly are more tolerant or less tolerant of pain.

Proprioception

There are sensory systems that send information to the central nervous system regarding the position and movement of limbs, the forces generated by skeletal muscles, and the position and motion of the body relative to the ground. This sensory function is referred to as proprioception, and the receptors involved are called proprioceptors. Widely distributed throughout the body, these include: muscle spindles that respond to skeletal muscle length, Golgi tendon organs that respond to skeletal muscle tension, receptors in the ligaments and capsule of joints that respond to limb position and movement, and receptors in the vestibular apparatus (semicircular canals, utricle and saccule) of the inner ear that respond to the position and movement of the head.

The influence of increasing age on proprioception is an understudied subject. However, there does appear to be deterioration at advanced ages in the ability to sense limb movement and to reproduce changes in limb position. There is a decrease in the number of receptors in the vestibular apparatus with advancing adult age, but initially this may be compensated for by changes in the response of the central nervous system to vestibular apparatus stimulation. Indeed, it appears that as an individual ages, he or she first become more responsive, and then with a further increase in age less responsive, to vestibular apparatus stimulation. The elderly often experience light-headedness and vertigo (the sensation that the person or the surroundings are spinning), and, at least in some individuals, altered functioning of the vestibular apparatus may be involved.

Hearing

Hearing refers to the response of the ear to sound and its perception by the central nervous system. Physically, sound is comprised of pressure waves propagated in some medium such as air, water, or wood. The pressure waves can be further characterized as having: 1. amplitude (i.e., the magnitude of the pressure change) which is perceived as loudness; 2. frequency (ranging from 20 to 20,000 waves per second) which is perceived as pitch, such as a note from a musical instrument; and 3. detailed wave structure, perceived as timbre, the quality of sound that distinguishes, for example, a violin and oboe playing the same note.

The anatomy of the ear is diagrammatically shown in Figure 6.5. Three components make up the ear: the outer ear, the middle ear, and the inner ear. The outer ear has an air-filled canal which ends at the ear drum (the tympanum). Sound waves travel through the air to the tympanum, which then moves in and out in a pattern determined by the characteristics of the sound waves. The movements of the tympanum are transmitted across the air-filled middle ear by three little bones (the malleus, incus, and stapes) that increase the force of the sound wave on the oval window of the liquid-filled inner ear. The inner ear has both a vestibular portion (shown in Figure 6.5 as the utricle and semicircular canal), which was discussed in the section on proprioception, and the cochlea, which will now be discussed. The cochlea has a sheet of tissue called the basilar membrane, which runs the length of this coiled, liquid-filled tube, and on this membrane are located hair cells that serve as the receptors for hearing. Each hair cell is attached to a nerve cell that carries nerve impulses, via the auditory nerve, to the central nervous system. The movements of the oval window cause pressure waves to be propagated in the liquid of the cochlea, and these waves stimulate the hair cells. The greater the magnitude of the pressure waves, the greater number of hair cells stimulated; this information is perceived by the brain as loudness. The frequency of the pressure waves influences both the frequency of nerve impulses originating from specific hair cells and which cells are maximally stimulated; this information is perceived by the brain as pitch. The spatial pattern of the hair cells being stimulated is perceived as timbre.

Hearing loss associated with senescence is called presbycusis, and it affects almost 40% of those 65 and older. With aging, many changes occur in the structures of the inner ear. There is atrophy, as well as loss, of the hair cells that encode high-frequency sound, and thus hearing loss. There is also a loss of hearing due to a loss of nerve cells of the auditory nerve with advancing age. In addition, hearing loss results from a reduced blood supply to the cochlea, as well as alteration in the structure of the basilar membrane. In the elderly, hearing loss due to changes in the external and middle ear appear to be of minor importance; although accumulation of wax

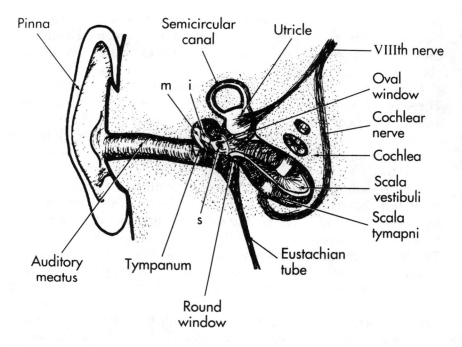

Figure 6.5 Diagrammatic section through the ear.

From "Neurophysiology," (3rd ed., p. 106), by R. H. S. Carpenter, 1996, London: Arnold. Copyright 1996 by Arnold. Reprinted with permission.

in the external ear does cause hearing loss in many, that problem can be easily rectified by removing the wax. Long-term exposure to intense sounds, such as those from power tools or loud music, are contributors to presbycusis. Some researchers feel that lifetime noise exposure, rather than intrinsic aging, is the cause of presbycusis.

As the result of presbycusis, many elderly have difficulty in distinguishing spoken words, a problem magnified by background noise. Also, in some individuals, presbycusis is a likely contributor to an age-associated decline in cognitive ability.

Vision

Light is a form of energy propagated by electromagnetic waves. Vision refers to the response of the eyes to a small part of the electromagnetic spectrum, specifically to wavelengths ranging from about 0.4 micrometers (blue light) to about 0.7 micrometers (red light); i.e., to what physicists refer to as visible light.

The visual system is composed of the eyes and neural structures, such as the optic nerves, and the neural pathways in the central nervous system, including the occipital cortex and other cortical regions. The eyes have two functional elements: the optical system that focuses light onto the retina; and the retina's photosensitive elements, i.e., the photoreceptors or rods and cones. A schematic diagram of the eye is shown in Figure 6.6. The optical system begins with the light entering the eye through the cornea and passing on through the aqueous humor; it then passes through the pupil (the opening in the iris) to the lens, and thence through the lens and the vitreous body to the retina. The lens is attached by ligaments to the ciliary muscles which can change the shape of the lens (making it more, or less, spherical), thereby enabling the light coming from various distances to be focused on the retina. (The closer an object, the more spherical the lens must be to focus its image on the retina.) The cones, which are in the central portion of the retina, are responsible for both high-acuity vision and color vision; they require relatively bright light. The rods, which are primarily in the peripheral parts of the retina, are sensitive to dim light. The rods and cones send messages (nerve impulses) to the central nervous system via the optic nerve and thence via central nervous system pathways to the cerebral cortex, where the images of objects are perceived

Changes occur in the visual system with age. Resting pupil size decreases, thus reducing the illumination of the retina. The ability of the lens to become more spherical when the person is looking at near objects (called the power of accommodation) progressively decreases with increasing age, and it is essentially completely lost in most people by age 60. This loss in the power of accommodation, a condition referred to as presbyopia, is the reason that most older people can not read a newspaper without corrective lenses. It is due to a change in the physical properties of the lens and in the functioning of the ciliary muscles. Visual acuity (the ability to see objects in fine detail) decreases with increasing age, even if the optical system deficits are rectified by corrective lenses. The loss in acuity does not appear to be due to the small loss of cones, but more likely results from a decrease in the number of neurons making up the optic nerve. The slight loss in rods that occurs with increasing age appears to be compensated for by hypertrophy of the remaining

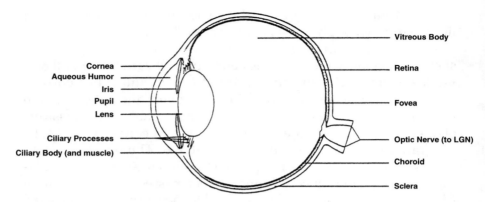

Figure 6.6 Cross-sectional diagram of the eye.

From "Vision," by C. T. Scialfa and D. W. Kline, 1996, in *Encyclopedia of Gerontology*, 2, p. 606, J. E. Birren (Ed.), San Diego: Academic Press. Copyright 1996 by Academic Press. Reprinted with permission.

rods. Nevertheless, there is impairment in rod vision in that the elderly have a reduced ability to adapt to very low-intensity light. The reduced ability of the elderly to discriminate colors in the green-blue-violet region of the visible light spectrum does not appear to be due to a defect in the cones, but rather relates to a yellowing of the lens with increasing age. Indeed, an alteration in the optical properties of the lens probably underlies the increased susceptibility to glare. Over 50, there is generally some loss in depth perception, for reasons that remain to be identified.

In addition to the changes just discussed, which occur almost universally with advancing age, there are several visual disorders that occur much more commonly in the elderly than in the young. Indeed, about two-thirds of those with severe visual impairment are over 65.

The most common of these are cataracts. A cataract is an opacity in the optical system of the eye, usually the lens; oxidative damage and glycation or glycoxidation appear to be causally involved. Cataracts diminish visual acuity and increase light scatter, resulting in an increase in image blur.

Glaucoma, a frequent problem for the elderly, is due to elevated intraocular pressure. Although both the production and drainage of the aqueous humor decrease with increasing age, drainage appears to be more affected than production, thus causing the intraocular pressure to rise. If untreated, glaucoma causes atrophic changes in both the optic nerve and retinal components that mediate peripheral vision, and this can result in tunnel vision.

Another not uncommon but poorly understood visual disorder is age-related macular degeneration. It is characterized by atrophy of cones and nerve cells within the central retina or macula. A rapid growth of new blood vessels results in fluid leakage into the macula, causing destruction of the cones. This markedly reduces visual acuity.

Taste and Smell

The sensory systems of taste and smell respond to chemicals coming to the organism from the external environment via the mouth and nose. Taste is a rather simple sense that involves four different sensations, salty, sweet, sour and bitter, each served by a specific kind of receptor. Primarily located on the tongue, these receptors are also present in other areas of the mouth and throat. Smell is a far more complex sense; many different receptors, located high in the nasal cavity, give rise to a broad spectrum of sensations. Indeed, much of what people refer to as taste is really smell, in that the mouth serves as a route for some odors to reach receptors in the nasal cavity.

Because of methodological difficulties, the effects of age on taste and smell are not well understood. Thus conclusions about these senses must be viewed as tentative. There appears to be no effect of age on the sweet and salty taste sensations, but there is some reduction in the bitter and sour taste sensations. Since smell sensations dim at advanced ages, many elderly live in a world of "pastel" food flavors.

Motor Functions

Earlier, in the discussion of neuromuscular function, it was pointed out that skeletal muscles contract when the muscle fibers receive nerve impulses from axons of motor neurons. We now turn to a broader consideration of the motor system, which refers to the complex neural systems that control the contractile activity of skeletal muscles in a fashion that permits complicated movements and maintenance of the body in space (i.e., posture). These tasks require groups of muscles to contract in appropriately timed sequences, and directing this process is one of the major functions of the nervous system. The processing of sensory input and other information by a complex network of synaptically connected neurons of the central nervous system is called central processing. The motor neurons are referred to as the final common pathway, because through central processing they receive neural inputs from various sensory systems and many different regions of the central nervous system, and they respond to this input by sending or not sending nerve impulses via their axons to skeletal muscle fibers.

Motor Responses to Sensory Input

Reaction time is defined as the time it takes to initiate a motor response following a stimulus. Reaction time has been found to be slowed at advanced ages. Although this is due, in small part, to the slowing of both muscle contraction and peripheral nerve impulse conduction velocity, it is primarily the result of the slowing of central processing. Reaction time in the elderly is slowed even more when the individual is confronted with a choice of alternative responses.

In the broad context, the elderly execute all movements more slowly than do the young, and the extent of this slowing increases as the movement increases in complexity. The good news is that this reduced speed appears to enable the elderly to maintain accuracy of movement. Nevertheless, the elderly do ultimately have some loss in their ability to precisely control skeletal muscle activity; this is illustrated in their handwriting, for example, or in their reaching for objects.

Thus the elderly have many different deficits in motor performance ability. These deficits are due to changes in central processing as well as changes in sensory and muscle function. This decrease in motor performance can interfere with daily living activities, but the extent to which it does varies greatly among individuals.

Posture and Balance

The alignment of the body parts in relation to each other is referred to as posture. Balance, which is closely related to posture, refers to the ability to maintain an upright position. The maintenance of posture and balance requires input from receptors of various sensory systems, including skin receptors, proprioceptors in the muscles, tendons and joints, receptors in the vestibular apparatus, and receptors in the visual system. The input from all these receptors must be rapidly processed and integrated by the central nervous system, primarily at a subconscious level, so as to modify motor nerve activities and thereby the contractile function of the appropriate muscles in a fashion that maintains posture and balance.

As discussed earlier, elderly people have deficits in these sensory systems, and thus would be expected to have postural and balance problems. However, an elderly person suffering from the deterioration of one sensory system does not necessarily have a problem, because these many systems provide redundancy of information. Nevertheless, both the speed of central processing and muscular strength do decrease with increasing adult age, and these factors tend to adversely affect posture and balance.

Balance is compromised to varying extents in the elderly. Postural sway when standing on two feet is somewhat greater in the old than in the young. However, the sway difference between the old and the young becomes much more pronounced when the individual stands on only one foot. Also, in tests such as maintaining balance when reaching for an object or opening a door, dynamic balance is shown to be compromised to varying extents in most of the elderly. The elderly are prone to falls, and deterioration of posture and balance is certainly contributory. It is important to note, however, that deterioration of posture and balance is only one of many factors underlying the increasing incidence of falls with advancing adult age.

Locomotion

Although walking requires little conscious attention, it is a complicated physiological process of skeletal muscle responses, which are involved in stride and support reactions, and postural and balance adjustments, as well as in coping with changes in the environment. A spectrum of sensory inputs also comes into play. Vision provides information on the direction and speed of movement; the vestibular system input assists in the maintenance of balance as well as signaling acceleration; and inputs from the proprioceptors in muscle, tendons, and joints provide information on the force of muscle contraction and the angle of the joints.

It would be expected that the elderly would show changes in walking because of the deterioration of the sense organs and joints and the loss in skeletal muscle strength. Surprisingly, there is only a modest change in walking in those who are free of discernible diseases, such as Parkinson's disease, strokes, cerebellar degeneration, and osteoarthritis. The most striking characteristic of walking in the healthy elderly is that it is slower than that of the young. This slowness is often not because they are not able to walk faster, but rather that they prefer to walk slower. Walking speed decreases more rapidly with increasing age in women than in men. During slow walking, the healthy old exhibit characteristics similar to those of the young, but during fast walking the elderly take shorter, more frequent steps than the young.

There are advantages for the elderly in their preference for a slow gait. Endurance of limb muscles is enhanced by shorter strides, and the energy cost of walking is reduced. A shorter stride length is also less taxing when ankle and knee joints are less flexible. In addition, in taking more steps to cover the same distance, both feet are on the ground for a greater fraction of the time; and this may be important for the elderly because of their more fragile balance. Also, a slow gait enhances the ability of the elderly to monitor the environment, thus enabling them to avoid hazards. Of course, there are situations when slow walking is not desirable, such as crossing an intersection governed by tightly timed traffic lights.

Age-Associated Motor Disorders

As mentioned above, many of the marked alterations in motor function (which most of us picture as typical of the elderly) are not seen in the healthy elderly, but are the result of age-associated disease processes. Several of these diseases were just mentioned, some of which have been or will be discussed in relation to other subjects (e.g., osteoarthritis was considered during the discussion of joints). In this section, two age-associated central nervous system motor disorders, Parkinson's disease and Huntington's disease, will be addressed.

The syndrome of Parkinson's disease includes tremors at near rest, rigidity (resistance to passive movement of limbs), slowness in initiating movements, deterioration of postural reflexes, lack of facial expression, and rapid, small steps with decreased associated movements such as arm swinging. In this disease, there is a loss of neurons of the substantia nigra that send axons to the striatum where they release the neurotransmitter dopamine. It is the lack of dopamine in the striatum that causes the syndrome of Parkinson's disease.

This is a progressive disease leading to severe disability and ending in death. The rate of progression varies among individuals, with death occurring within 5 years in 25% and within 10 years in 60% of the cases. It is estimated that more than 150,000 people in the United States suffer from Parkinson's disease. Onset is rarely before age 40, and its incidence increases with increasing age, with 68 years the average age of onset.

Huntington's disease is a relatively rare genetic disorder characterized by age-associated motor dysfunction. The average age of onset is 35 to 42 years, and it progresses in severity until death occurs in about 15 years. The problem stems from deterioration of neurons in the corpus striatum, resulting in uncontrollable and jerky movements. Further progression of the disease involves memory loss and dementia. This disease is of particular interest to gerontologists because it appears to be an example of Medawar's concept of late-acting detrimental mutations as a cause of senescence.

Autonomic Nervous System

The autonomic nervous system regulates the functioning of most of the organ systems of the body. There are two separate branches of the autonomic nervous system: the sympathetic nervous system and the parasympathetic nervous system. The sympathetic nervous system plays a major role in coping with dangerous or stressful situations, while the parasympathetic nervous system functions in a restorative role, often by countering the action of the sympathetic nervous system.

The sympathetic nervous system originates in the thoracic and lumbar regions of the spinal cord. It sends nerve fibers to target cells (smooth and cardiac muscle cells and gland cells) throughout the body. These nerve fibers, when stimulated, release the neurotransmitter norepinephrine at these target cells; it is the interaction between norepinephrine and receptors of the target cells that modulates the functioning of the target cells. The adrenal medulla is essentially a component of the sympathetic nervous system; when stimulated, it secretes into the blood a mix of norepinephrine and epinephrine (the latter commonly referred to as adrenaline), which is carried by the cardiovascular system to the target sites where they modulate the functioning of the target cells.

The parasympathetic nervous system originates in the brain stem and the sacral region of the spinal cord. It sends nerve fibers to heart and smooth muscle cells and gland cells throughout the body and, when stimulated, these nerve fibers release the neurotransmitter acetylcholine at these target cells. The acetylcholine interacts with receptors on the target cells to modulate the functioning of the target cells.

The functioning of the autonomic nervous system is influenced by input from the sensory systems and forebrain regions, including the cerebral cortex. The hypothalamus is the major site of the central processing of this broad array of information. Thus, the activity of the autonomic nervous system is modified so as to assist the organism in coping with the various situations it encounters.

Few generalizations can be made regarding the influence of aging on the functioning of the autonomic system. Therefore, age effects will be considered in the context of specific organs or organ systems. It will be important then to assess the information within the general framework of the autonomic nervous system just presented. However, there is one general aspect that should be discussed now. The concentration of norepinephrine in the blood increases with increasing age, and the increase in its concentration in response to stress is greater and lasts longer in the elderly. The level of blood norepinephrine reflects the activity of the sympathetic nervous system and, to a lesser extent, the adrenal medulla. These facts strongly indicate that with increasing age, there is an increase in sympathetic nervous activity. However, this does not necessarily mean that the elderly suffer from excessive sympathetic nervous system activity, because there is also evidence that, at least in some organs, there is a reduced ability of target cells to respond to norepinephrine.

Sleep

The electroencephalogram or EEG has played a major role in the study of sleep. Placing large electrodes on the scalp enables the recording of the average electrical activity of a very large number of cortical neurons, and the record of this activity is called an EEG. In one state of sleep, the electrical activity changes to a lower wave frequency and a higher voltage than that recorded when awake. During this state, called slow wave sleep (SWS), muscular tone is much decreased. However, SWS is interrupted about every 2 hours by an electrical pattern that resembles the awake state, though it is clear that the person is not awake. These interruptions are often referred to as paradoxical sleep; they are also called rapid eye movement, or REM, sleep because they are characterized by increased bodily movements, including eye movements.

With increasing age, the most marked change in the EEG is the reduced voltage of the electrical potentials during SWS. Also, the length of time of REM sleep is decreased, as are eye movements. In the elderly, sleep is fragmented with frequent awakenings. A major reason for the awakenings is an age-associated increase in sleep apnea (defined as a cessation of breathing for 10 seconds or longer). In older men, sleep is also interrupted because of urinary urgency that is secondary to enlargement of the prostate. As a result of these problems, more than one-third of people over 60 complain of sleep disturbances.

Cognitive Functions

Cognition refers to processes of the mind such as perceiving, remembering, thinking, learning, and creating. Although the number of functional domains embraced by cognition is great, our discussion will be limited to a few of the major ones. There are many age-associated disorders that result in a marked decline in cognitive processes, and they will be presented at the end of this section. However, our discussion will begin with what happens to cognition in those people who grow old without discernible brain disease.

Attention

The ability to focus on and perform a simple task without losing track of the task objective does not undergo an appreciable age-associated change. Since this functional activity primarily involves neuronal circuitry of the brain stem and thalamus, it appears that these brain regions remain functionally intact. However, in the presence of distractions, the elderly are less able to focus on a task than are the young. While laboratory studies have identified this attention deficit, the extent to which it affects everyday performance of the elderly remains to be determined.

Memory

As people grow older, most of them complain of memory loss. However, memory is a complex phenomenon, and not all aspects of memory deteriorate with advancing age. Unfortunately, the neurophysiological basis of memory is not well understood. However, theoretical models of memory provide some insights. The model that will be used here is called the *Memory System Theory*, which views memory as a three-component interacting system.

The first component is called sensory-perceptual memory, and involves the initial processing of sensory information. Although there may be some deterioration in this processing, it does not appear to be a major reason for memory deterioration with increasing adult age.

The second component is called short-term memory, and it is characterized by conscious awareness. It consists of information currently in the mind; this is a combination of information coming from the first component and that being retrieved from the third component (i.e., long-term memory). Short-term memory is further divided into primary memory and working memory. Primary memory refers to how many things a person can keep in mind at one time, an example being the longest string of numbers a person can repeat without making an error. There appears to be little, if any, change in primary memory with increasing adult age. Working memory refers to keeping information in mind while engaging in another task. An example would be remembering the last word in a series of sentences while engaged in remembering the meaning of each sentence. Working memory has been found to deteriorate at advanced ages.

The third component, long-term memory, is information that is no longer kept in conscious awareness (i.e., in the second component) but must be retrieved as needed. Long-term memory is divided into episodic memory, semantic memory, and procedural memory.

Episodic memory involves recollections that are actively retrieved within the context of a prior personal experience, such as where the keys to the automobile were left, what an individual has been requested to purchase at a pharmacy, or when a particular medication should be taken. The elderly have greater problems with episodic memory than do younger people.

Semantic memory involves the retrieval of conceptual information. Examples are the ability to define words or to name the authors of particular books. Semantic memory is well retained in the healthy elderly, although it may take longer to retrieve this information than is the case in younger people.

Procedural memory refers to a recollection that does not require conscious awareness. Examples are typing and bicycle riding, which require motor and cognitive skills but, once acquired, are automatic. This kind of memory does not deteriorate with age.

Intellectual Functions

With respect to semantic knowledge (verbal ability in vocabulary, information, and comprehension), intellectual performance changes little in the absence of disease during adult life, at least until the middle of the ninth decade. Indeed, in some studies, certain aspects of intellectual function (e.g., vocabulary, information and practical judgment) were actually found to improve in healthy individuals between the third and seventh or eighth decades of life. However, timed tasks (e.g., the speed of addition or subtraction) show marked deterioration at advanced ages.

Although old people can learn, the speed of learning decreases with advancing age. Moreover, some kinds of learning become increasingly difficult and ultimately impossible with increasing age, e.g., tasks requiring great perceptual speed and a high level of physical coordination. However, the elderly can master most new tasks, provided that they are allowed to learn at their own pace.

Some elderly individuals maintain a high level of creativity into the tenth decade of life (e.g., Pablo Picasso, Pablo Casals, and Frank Lloyd Wright), and some even develop creativity at that age (e.g., Mary Robertson "Grandma" Moses). However, because of the difficulty of defining creativity, it is not known how common this phenomenon may be.

Age-Associated Cognitive Disorders

Dementia is defined as a decline in intellectual functions and a deterioration in personality and emotions; it is a major age-associated syndrome. The prevalence of dementia increases with age, as does its incidence. The prevalence is in the order of 1% in the age range of 65 to 70, about 10% in the age range of 80 to 85, and about 40% in the age range of 90 to 95. While more than 70 disorders may produce dementia, the most common causes are Alzheimer's disease and cerebrovascular disease.

Alzheimer's disease begins in an insidious manner, for it can be difficult to distinguish from mere forgetfulness, but the symptoms progress in severity, with the individual remaining alert and awake until the terminal stages. Memory disturbance is the most prominent initial symptom. With further progression in severity, there are: a decrease in ability to express oneself in speech and writing and in the ability to understand written and spoken language; a loss in ability to copy simple drawings; and a reduced ability to recognize familiar objects by sight. Ultimately, all mental functions, including emotional ones, are impaired.

Alzheimer's disease involves degeneration of neurons in the cerebral cortex and, in particular, the hippocampus. Postmortem pathological analysis of the affected brain regions reveals what are called senile plaques and neurofibrillary tangles. The senile plaques contain a core of protein, known as beta-amyloid, surrounded by swollen degenerating nerve terminals and two classes of glial cells, astrocytes and microglial cells. These plaques are one of the hallmarks of Alzheimer's disease in that they are not found in most other neurodegenerative diseases, although they are seen in Down's syndrome and, in small numbers, in the brains of what appear to have been cognitively normal elderly. The other hallmark is the neurofibrillary tangles found inside the axons and dendrites of brain neurons. There is strong, but not unequivocal, evidence that the beta-amyloid protein deposited in the plaques is toxic and plays a causal role in the genesis of Alzheimer's disease.

A very small fraction of Alzheimer's cases is early-onset (occurring between ages 28 and 65) and familial (i.e., clearly inherited). Genetic studies have revealed that all of these patients do not have the same genetic defect. Indeed, as of now, three different groups of such patients have been identified, each with a different gene underlying the disease; it is likely that further research will identify patient groups in whom still other genes are involved. Hopefully, the careful study of the groups of patients with early-onset, familial Alzheimer's disease will provide insights into the much more common late-onset form of the disease. Indeed, the studies that have been conducted with the early-onset, familial Alzheimer patients further imply a causal role of the beta-amyloid protein in this disease.

There appears to be a genetic predisposition for late-onset Alzheimer's disease. The epsilon-4 allele of the apolipoprotein E gene has been found to increase the risk of the disease at advanced ages. When one of the homologous chromosomes has this allele (about 23% of the population), the risk of developing Alzheimer's disease at 60-plus years is 2 to 4 times greater than those without the allele. When both of the homologous chromosomes have this allele (about

3% of the population), the risk of developing Alzheimer's disease is 5 to 9 times greater. It is important to note, however, that there are people who have reached 100 years of age with the allele in both homologous chromosomes who do not suffer from Alzheimer's disease. Very recent evidence suggests that a common allele of alpha$_2$-macroglobulin gene (about 30% of the population) may also be linked to the common form of Alzheimer's disease.

Psychosocial factors are also associated with the development of Alzheimer's disease, particularly the late-onset varieties. Of these, the best studied is educational level, which has been investigated in the United States, Europe, and China. The incidence of Alzheimer's disease is two- to three-fold greater for those with little or no education compared to those with high school or greater education.

An effective therapy for treatment of Alzheimer's patients has yet to be developed, but some preventive measures appear to hold promise. Estrogen replacement in postmenopausal women may delay or prevent the occurrence of Alzheimer's disease. Also, nonsteroidal anti-inflammatory drugs, such as aspirin and ibuprofen, taken for prolonged periods (e.g., for arthritic conditions) also seem to offer protection; further research, however, is needed to be certain of this preventive action. There is also suggestive, but very preliminary, evidence that folic acid supplementation may be effective in this regard

The other major form of dementia is secondary to cerebrovascular disorders, i.e., problems with the blood flow to the brain. Of these, the best known is multi-infarct dementia, which is due to multiple infarcts (obstructions of blood vessels due to a blood clots), many too small to have produced a major clinical incident. There is often a history of either strokes or transient ischemic attacks (TIAs). Typically, there is a sudden appearance of dementia with stepwise deterioration, though often with a fluctuating course. However, in some cases, the onset of dementia may be gradual with a progressive increase in severity. Major risk factors are high blood pressure, diabetes mellitus, and smoking. Effective treatment of high blood pressure reduces the risk of this form of dementia.

Recently another form of cerebrovascular-related dementia has been recognized; it is referred to as white matter lesions. This dementia is thought to be secondary to a narrowing of the lumen of the small arteries and arterioles (minute arteries) that nourish the white matter of the brain.

A mix of Alzheimer's disease and dementia due to cerebrovascular disease is increasingly being recognized. Indeed, such a mix may be the most common form of dementia.

CARDIOVASCULAR SYSTEM

The cardiovascular system circulates the blood throughout the body. Simply stated, it is composed of a pump (the heart) and tubes (the blood vessels). Figure 6.7 schematically depicts the cardiovascular system. The system has two circuits in series: One circuit (the pulmonary circuit) circulates the blood through the lungs; and the other circuit (the systemic circuit) circulates the blood through the rest of the body. The heart is a four-chambered structure comprised of blood-filled cavities whose walls are made up primarily of heart muscle cells. In Figure 6.7, it is depicted, for illustration purposes, as a left heart and a right heart. In reality, the heart has a left and right side, each with two chambers in series: The left atrium has a relatively minor pumping function that helps fill the left ventricle with blood; the left ventricle, a powerful pump, then propels the blood through the systemic circuit. The right atrium has a minor pumping function that helps fill the right ventricle with blood; the right ventricle then pumps the blood through the pulmonary circuit.

In a resting person, the left ventricle pumps about 5,000 ml of blood per minute into a large artery called the aorta, whence it flows through the rest of the systemic arteries. The blood in the systemic arteries has a high pressure and flows from the arteries through minute arterial vessels called arterioles to the capillaries of the tissues and organs of the body (arterioles and capillaries are not shown in Figure 6.7). The capillaries are the major site of exchange of materials (such as nutrients, waste products, and hormones) between the blood and the cells of the tissues and organs. The blood is collected from the systemic capillaries by the systemic venous system. While the pressure in the veins is low compared to the arteries, it is greater than that of the right atrium. Because of this pressure difference, the blood flows from the veins to the right atrium and thence to the right ventricle, which pumps it into the pulmonary arterial system. Although the blood pressure in the pulmonary arteries is low compared to the systemic arteries, it is sufficient to propel the blood through the pulmonary capillaries

The Human Aging Phenotype

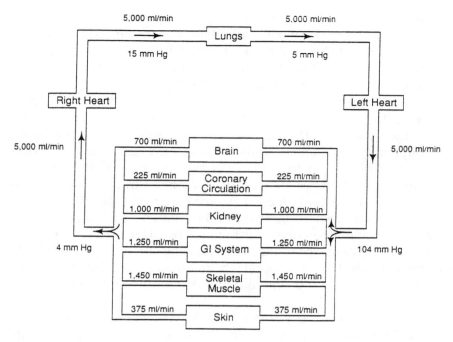

Figure 6.7 A schematic diagram of the human cardiovascular system including blood flows and pressures typically found in people at rest.

From "Essential Medical Physiology," p. 148, L. R. Johnson (Ed.), 1992, New York: Raven Press. Copyright 1992 by Lippincott-Raven Press. Reprinted with permission.

and thence the pulmonary veins to the left atrium and finally the left ventricle.

Heart

In healthy people, there is a modest increase in the size of the heart from ages 20 to 80. This increase is due primarily to an increase in thickness of the wall of the left ventricle of the heart, resulting from hypertrophy of the wall's cardiac muscle cells. The age-associated increase in heart mass is far greater in people who suffer from high blood pressure (hypertension).

Conductile System

In addition to cardiac muscle cells that serve as the motor of the pump, the heart also has specialized muscle cells called conductile cells that initiate and coordinate the pump function. A group of these cells, called the sinoatrial (SA) node, is located where the systemic venous system enters the right atrium. These cells spontaneously generate action potentials (although this activity is usually modified by nerve fibers supplying the SA node), and these action potentials spread to atrial muscle cells and thence to a group of conductile cells (collectively called the atrioventricular or AV node) near the ventricles. The AV node, in turn, sends action potentials through a specialized pathway of conductile cells that cause ventricular muscle cells to generate action potentials. This ordered sequence of action potentials causes the heart muscle cells to contract in an orderly and timed sequence for effective functioning as a pump. An electrocardiogram (EKG) provides a record of the conduction of action potentials through the heart, and it is obtained by the proper placement of electrodes on the surface of the body. In normal people, the SA node sets the frequency of the heart beat (i.e., the heart rate), and thus the cells of the SA node are called the pacemaker cells.

After age 60, there is a progressive decrease in the number of cells in the SA node. There is also some decrease in the resting heart rate with increasing adult age, and this appears to be due, in part, to a change in the SA node's pacemaker function. There is also some change with age in the AV node and its connection to the conductile system in the ventricles, causing a minor delay in the progression of action potentials from the atria to the ventricles. Increasingly common with increasing age are abnormal rhythms (arrhythmias) of the heart, such as a too rapid (tachycardia) or too slow (bradycardia) heart rates, or the occurrence of pacemaker cells at sites other than the SA node. It has been suggested that sudden death due to an arrhythmia may be relatively common in the very old.

Pump Function

The heart goes through contraction (systole) and relaxation (diastole) some 60 to 100 times per minute in a healthy person at rest. During a systole, a volume of blood (the stroke volume) is ejected from the cavity of the left ventricle into the arteries of the systemic circuit, and on average a similar volume of blood is ejected into the arteries of the pulmonary circuit by the right ventricle. The total amount of blood pumped into the systemic circuit by the left ventricle in one minute is called the cardiac output; in a healthy person at rest, it is about 5,000 ml per minute.

In discussing age changes in pump function, the focus will be on the left ventricle because it has the enormous task of supplying blood to every part of the body except the lungs. (The pulmonary circuit will be discussed further when the respiratory system is presented.) For now, a good starting point is when, in a person at rest, the left ventricle goes into diastole and the mitral valve (the valve between the left atrium and left ventricle) opens. At this time, there is a rapid flow of blood from the pulmonary venous system through the left atrium to the left ventricle, because the pressure in the ventricle is lower than that in the pulmonary veins. With aging, there is a slowing of the flow of blood into the left ventricle during early diastole, but this is compensated for by the increased amount of blood pumped by left atrial contraction in late left ventricular diastole. Thus, at rest, the total amount of blood entering the left ventricle during diastole is similar for old and young people of the same size and gender.

As the left ventricle goes into systole, the pressure within its chamber increases, thereby rapidly closing the mitral valve. With a further increase in pressure within the left ventricular chamber, the aortic valve (between the left ventricle and the aorta) opens, and the left ventricle propels blood into the aorta, the start of the systemic circuit. The stroke volume in resting healthy people is similar for the old and young of the same size and gender, as is the cardiac output. However, in the healthy young and old, there is one difference in the pump function of the left ventricle. The contraction of left ventricular muscle cells is prolonged with increasing age, and this prolongation helps the healthy old maintain a stroke volume similar to that of the young.

Although there are only small age changes in the functioning of the heart as a pump in healthy people at rest, substantial differences emerge when a person is challenged. However, to discuss these differences, it is first necessary to consider the neural regulation of heart function.

Neural Regulation

The autonomic nervous system plays a major role in the regulation of heart function. The parasympathetic system sends nerve fibers to the SA node, where they release acetylcholine to decrease the heart rate. The sympathetic nervous system sends nerve fibers to both the SA node, where they release norepinephrine to increase the heart rate, and to the ventricular muscle cells, where norepinephrine increases the force of contraction. There appears to be no major age-associated change in the influence of the parasympathetic nervous system on the heart.

The influence of the sympathetic nervous system, however, is blunted with increasing age. Exercise is a good example of how this age-associated blunting alters the pump function of the heart in response to a challenge. In young people, the need to increase the cardiac output during exercise is met by an increase in the activity of the sympathetic nerve fibers to the heart, which increases heart rate and stroke volume, the latter because of the increased contractility of the ventricular cardiac muscle cells. In healthy old people, there is much less increase in heart rate and ventricular contractility in response to exercise. Instead, there is an increase in the blood volume in the chamber of the left ventricle at the end of diastole, causing an increase in the length of the left ventricular muscle cells. Within limits, an increase in the length of cardiac muscle cells increases the force of their contraction. Therefore, in the elderly, there is an increase in stroke volume during exercise that is secondary to the increased diastolic volume of the left ventricle. Thus the healthy old are able to increase cardiac output in response to exercise, but by a different mechanism than that of the young. However, it is important to note that in the elderly who suffer from age-associated cardiovascular disorders, this compensatory ability is compromised.

Coronary Heart Disease

A major age-associated medical problem is cardiac ischemia, an inadequate supply of oxygen to the heart muscle. At first glance, this would seem to be a paradox, because all of the blood flow of the body goes through the chambers of the heart. However, the walls of the heart are too thick to permit blood in the chambers to supply an appreciable amount of oxygen (and other nutrients) to the heart muscle. Rather, blood for the heart itself must be supplied by the coronary circulation, which begins with the coronary arteries that branch off the aorta near the heart, and which provides the heart muscle with capillaries where oxygen and fuels diffuse to the heart muscle cells and metabolic waste products enter the blood. The coronary arterial system can be subject to atherosclerosis, a process involving the development of plaques that narrow the lumen of the arteries. If this narrowing is sufficiently great, it causes the heart muscle to suffer from ischemia (an inadequate blood supply), which can lead to death of heart muscle cells, referred to as myocardial infarction. An infarction can be sudden when, in addition to an atherosclerotic plaque, it involves a thrombus (formation of a blood clot in a coronary artery) or an embolus (a blood clot that has detached from another region of the body and lodges in a coronary artery).

Both the incidence and prevalence of coronary heart disease increase with increasing age. In the age range of 65 to 75, there is a 50% prevalence, and a 60% prevalence in those over 75. The major risk factors include: elevated systolic blood pressure; high levels of serum LDL (low density lipoproteins) and low levels of serum HDL (high density lipoproteins); left ventricular hypertrophy; diabetes mellitus; elevated plasma glucose levels; and smoking. Although in healthy people, there are only modest changes in heart pump function due to aging, coronary heart disease can cause serious deficits that range from difficulty in exercising to decreased function at rest. Coronary heart disease is a major contributor to death of the elderly.

Congestive Heart Failure

Heart failure is not a disease in itself but a syndrome, that is, a complex of signs and symptoms. It involves a failure in the pump function of the heart, which results in inadequate oxygen delivery to the tissues, or pulmonary congestion, or systemic venous congestion, or all three. It is an age-associated syndrome in that 75% of the patients suffering from congestive heart failure are over 60. The two major causes of this syndrome are coronary heart disease and hypertension (high blood pressure); the latter will be discussed presently. Lesions in the heart valves, not an uncommon problem in the elderly, may also be a cause.

Arteries

Our discussion will focus on the systemic arteries, vessels whose blood pressure is high compared to the pressure in the other vessels of the systemic circuit. These vessels are overfilled with blood and are thus continuously stretched by the pressure of the blood.

Arterial Structure

With increasing age, there is an increase in the diameter of the lumen (the blood-filled channel) of the large arteries. The walls of these arteries increase in thickness, and they also become stiffer. While there is less increase in the lumen diameter of the smaller peripheral arteries, there is greater increase in their wall thickness. These age changes in arterial structure are due to several factors: a decrease in elastin relative to collagen in the arterial walls; an increased mineralization of the elastin with calcium and phosphorus; and an increase in sustained contractile activity of smooth muscle in the walls of the arteries.

The ejection of blood during systole from the left ventricle into the aorta causes a pressure wave to be transmitted along the walls of the arteries, which is called a pulse wave or, more commonly, the pulse. One of the hallmarks of aging of the cardiovascular system is the increased velocity of the pulse wave, which stems from the increased stiffness of the arterial walls.

Arterial impedance refers to the opposition to the ejection of blood into the aorta from the left ventricle. An increase in the diameter of the lumen of the large arteries decreases impedance, while an increase in stiffness in arterial walls increases it. Through middle age, these two factors are in balance and impedance does not change appreciably, but at advanced ages the increase in stiffness prevails and impedance increases.

With increasing age, atherosclerosis progressively alters the structure of arteries in many, but not all, people. Atherosclerotic plaques can impede blood flow in the arteries in which they occur, particularly by serving as sites of blood clot formation. The consequences of atherosclerosis have already been discussed in regard to the coronary circulation. In addition, atherosclerotic plaques commonly occur in the internal carotid arteries near their origin in the neck, the middle cerebral arteries, the vertebral arteries, and the basilar arteries; as these plaques grow in size with increasing age, they often become sites for formation of blood clots that cut off the blood supply to regions of the brain.

Blood Pressure

When the term blood pressure is used without stating the region of the cardiovascular system under consideration, it refers to the blood pressure in the systemic arteries. The systemic arterial blood pressure changes continuously, because blood is pumped into the arteries periodically by left ventricular systole and continuously flows out of the arteries through the arterioles into the capillaries. For each beat of the heart, the highest pressure (the systolic pressure) occurs near the end of left ventricular systole, and the lowest pressure (the diastolic pressure) occurs at the beginning of left ventricular systole, just before the aortic valve opens. Thus when a physician tells a patient his or her blood pressure is 120 over 80, the higher number refers to the systolic pressure and the lower to the diastolic pressure. Physiologists also like to consider the mean pressure (i.e., the average pressure during a heart beat) because conceptually that can be considered the sustained force driving the blood through the resistance vessels (arterioles) into the capillaries.

Several population studies have shown that systolic, diastolic and mean blood pressures increase between the ages of 20 and 70. The increase in mean and diastolic blood pressure is primarily due to an increase in the resistance of arterioles. The increase in systolic pressure stems from the increased stiffness of the walls of the large arteries as well as to increased resistance of the arterioles. However, not all populations show age-associated increases in blood pressure. For example, a study of 144 Italian nuns showed no significant increase in blood pressure with increasing age. Suffice it to say, body weight, physical exercise, and smoking have been shown to modify the age-associated increase in blood pressure.

Hypertension, persistently elevated blood pressure, occurs in over 60% of the elderly. There are two basic forms. One involves elevation of both the systolic and diastolic blood pressures, defined numerically as a systolic pressure of 140 mmHg and above and a diastolic pressure of 90 mmHg and above. The other form, termed Isolated Systolic Hypertension, is defined as a systolic blood pressure greater than 160 mmHg and a diastolic pressure of 90 mmHg or less. Both forms of hypertension are dangerous in that they increase the risk of stroke or heart attacks, but fortunately both are treatable by diet, exercise, medication, or a combination thereof.

About 30% of the population over 65 show postural hypotension (low blood pressure); i.e., upon standing, their systolic blood pressure drops by 20 mm Hg and remains at that reduced level for at least 1 minute. By decreasing blood flow to the head, postural hypotension is a contributor to falls by the elderly. This hypotension is caused by altered reflex responses to falling blood pressure, the most important being a blunting of the arterial baroreceptor reflex, which readjusts blood pressure by modifying both heart rate and resistance of the arterioles. The elderly are also prone to postprandial hypotension (i.e., a fall in blood pressure an hour or so after eating); this is due to an inability to compensate for the decrease in the resistance of the arterioles of the gastrointestinal tract by increasing the resistance of the arterioles of other regions.

Microvasculature

The fine vessels that connect the arteries to the veins are termed the microvasculature. Starting on the arterial side, the vessels involved are the arterioles, metarterioles, capillaries, and the venules which deliver the blood to the veins.

Resistance Vessels

The arterioles are the major site of resistance to blood flow. These minute vessels have a thick layer of smooth muscle in their walls, and contraction of this muscle (regulated by the autonomic nervous system) reduces the lumen diameter and thus increases resistance to blood flow, while relaxation of the muscle has the opposite action. By controlling the contraction of the smooth muscle, and thus the diameter of the arterioles, the autonomic nervous system plays a major role in regulating arterial blood pressure. In addition, the autonomic nervous system is able to increase the diameter of arterioles in one region while decreasing the diameter in another region and, in this way, redistributing blood flow. For example, during exercise the arterioles supplying the exercising muscles are dilated while those supplying the visceral organs are constricted, thereby diverting blood flow to the exercising muscles.

Changes in the functioning of the arterioles may well play a major role in the tendency to develop hypertension with increasing age. In many people, there is an age-associated increase in resistance to blood flow because the arterioles are too constricted, and this appears to be a major reason arterial blood pressure tends to increase with age.

Also, the increased occurrence of postural and postprandial hypotension at advanced ages may involve altered responsiveness of arterioles, in addition to changes in the functioning of the nervous system. However, whatever the reason, it is clear that for many elderly, the regulation of the size of the lumen of the arterioles by the autonomic nervous system is no longer fully effective in redistributing blood flow.

Exchange Vessels

The capillaries (tiny tubes with very thin walls) are the site of exchange of materials between the blood and cells of the body. The blood flow through capillaries is controlled by: 1. the volume of blood that arterioles allow to flow to that region, and 2. the precapillary sphincters. Each of the latter is a thin band of smooth muscle located at the point where the capillary branches off the metarteriole. Basically, these sphincters are either relaxed, thereby letting blood flow into the capillary, or they are contracted, thus preventing blood from entering the capillary.

The influence of aging on capillaries is an under-studied subject. The limited information available indicates no change with age in their structure and function. However, in some tissues, the number of capillaries decreases at advanced ages.

Venous System

The blood flows from the capillaries into the venules and thence to the veins of the systemic circuit. All veins are thin-walled compared to the arteries, and contain smooth muscle that is controlled by the sympathetic nervous system. The blood pressure in the veins is low, just a few mm Hg. There are valves in the veins that function so as to direct flow from the capillaries to the right atrium. Most of the blood in the systemic circuit is in the venous component of the system, and contraction of the smooth muscle of the veins can redistribute significant volumes of blood within the vascular system.

With increasing age, there is a reduction in the distensibility of the veins and in the strength and speed of smooth muscle function in the veins. There may also be some loss in the efficiency of the sympathetic nervous system control of the smooth muscle of the veins. In addition, there is an age-associated widening of the veins; this interferes with the proper functioning of the valves and, as a result, there tends to be a collection of fluid (an example of edema) in the legs.

RESPIRATORY SYSTEM

The function of the respiratory system is to supply the cells of the body with oxygen to meet their metabolic needs and to remove from the body the carbon dioxide generated by the metabolic activity of the cells. To do this requires a pump system to move air in and out of the lungs, plus a transport system to move 1) the O_2 from the lungs to the cells of the body and 2) the CO_2 from the cells of the body to the lungs.

Lung-Thorax Pump

The lungs are basically air-filled elastic bags with an opening to the atmosphere via the mouth, throat, and trachea (windpipe). A thin film of liquid, called pleural fluid, serves to adhere the surface of the lungs to the inner surface of the chest wall, including that of the diaphragm. This adherence prevents the lungs from separating from the chest wall when the volume of the thorax expands during breathing. Thus the lungs and chest wall move as a single unit.

The respiratory tract, which includes the lungs, is a complex tubular system. The trachea divides into two airway tubes, called the bronchi, which in turn divide and so on until there are 20 to 23 such divisions of ever smaller-diameter tubes (bronchioles), ending in 1 to 8 million terminal tubes. Numerous tiny air sacs, called alveoli, sprout from each of these terminal tubes, resulting in about 300 million alveoli. Each of the air sacs or alveoli is in close proximity to the blood in a capillary of the pulmonary circuit.

When the respiratory skeletal muscles of the chest wall and the diaphragm are relaxed, the volume of air in the lungs is called the functional residual capacity (FRC). The FRC is determined by the elastic forces of the lungs tending to reduce the volume of air in the lungs and the elastic forces of the thoracic wall and diaphragm tending to increase it. With aging, there is a decrease in both these forces, though the decrease in force is much greater for the lungs. As a result, the FRC increases with increasing age; this is the reason that elderly people tend to develop what is called a barrel chest.

Inspiration (breathing in) causes atmospheric air to flow into the respiratory tract, including the lungs. The driving force for inspiration is the contraction of the diaphragm and inspiratory skeletal muscles of the thoracic wall; this action expands the volume of the thorax and thus the lungs, thereby reducing air pressure in the lungs below that of the atmosphere. Upon inspiration, air from the atmosphere flows into the trachea and thence through the complex of airway tubes into the alveoli, until the air pressure in the alveoli is the same as that of the atmosphere. When contraction of the inspiratory muscles ceases, air is expelled from the lungs (expiration) because the elastic recoil of the lungs decreases the volume of the thorax and lungs, thereby causing the air pressure in the lungs to be above that of the atmosphere. Thus air flows out of the alveoli, through the complex of airway tubes into the atmosphere, until the air pressure in the alveoli is the same as that in the atmosphere. The volume of air in the lungs can be even further decreased, below that due to elastic recoil, by contraction of the expiratory muscles of the chest wall; this is referred to as forced expiration.

The contraction and relaxation of the inspiratory and expiratory muscles are controlled by neurons located in the brain stem. A spectrum of sensory and volitional inputs regulates the activity of these neurons and thus the frequency and depth of breathing.

The functioning of the lung-thorax pump, referred to as pulmonary function, can be tested in various ways. A commonly used measurement is the vital capacity, which refers to the maximum volume of air that can be expired after a maximum inspiration. The vital capacity decreases with increasing age, because of the changes in the physical properties of the thorax-lung system and the diminished force-generating ability of the respiratory muscles. The residual volume (the amount of air remaining in the lungs after a maximal forced expiration) increases with increasing age for the same reasons. Another widely used and informative test is the measurement of the volume of air exhaled during the first second of a forced expiration following a maximum inspiration; this test is referred to as the FEV_1. There is a progressive decrease in the FEV_1 after age 25, because of the increased resistance to air flow in the bronchioles, the change in elastic properties of the lungs, and the decrease in the force generated by the respiratory muscles. The decline in FEV_1 is about 30 ml per year for nonsmokers, and it is greater than that for smokers.

The volume of air that the lung-thorax pump brings to and removes from the alveoli per minute is called the alveolar ventilation; in the resting adult, it is about 4 liters per minute. While there is always an unevenness in the ventilation of the approximately 300 million alveoli, with some being overventilated (i.e., receiving more air than needed to oxygenate the pulmonary capillary blood) and others being underventilated, the unevenness increases with increasing age. In spite of the age changes in the lung-thorax pump, alveolar ventilation in the healthy elderly is not sufficiently altered to limit their ability to carry out vigorous exercise. However, such is not the case for those suffering from chronic obstructive pulmonary disease. In such individuals, there is a much more marked decrease with age in the FEV_1, and limitations in alveolar ventilation in these individuals can result in serious disability. Fortunately, this condition rarely occurs in nonsmokers. Moreover, it only occurs in a subset of smokers, indicating that in addition to aging and smoking, a genetic component is involved.

Gas Transport Between Alveoli and the Tissues

Oxygen in the alveolar air diffuses into and dissolves in the fluid of the alveolar wall and then diffuses into the blood flowing in the capillaries of the pulmonary circulation. Most of the oxygen does not remain in the dissolved form, but rather it reacts with the hemoglobin in the red blood cells to form oxyhemoglobin. It is in this form that most of the oxygen is transported by the cardiovascular system to the tissues of the body.

Two age changes affect the diffusion of oxygen from the alveolar air into the blood. One is the age-associated decrease in the surface area of the alveoli. The other is the age-associated increase in the unevenness of alveolar ventilation. Indeed, the concentration of dissolved oxygen in the arterial blood (measured as the partial pressure of oxygen) decreases with age at the rate of about 4% per decade after age 20. Fortunately, this decrease has little effect on the amount of oxygen carried to the tissues because hemoglobin is almost fully loaded with oxygen, even at the lower concentrations of dissolved oxygen that occur in the healthy elderly. However, a further decrease in dissolved oxygen can cause problems, and thus the elderly are more vulnerable to pulmonary diseases, such as pneumonia.

Of course, a decrease in the amount of hemoglobin in the blood (i.e., anemia) would markedly reduce the amount of oxygen carried in the blood. This does not occur in the healthy elderly, but is a frequent result of age-associated diseases.

As the blood flows through the capillaries of the systemic circuit, much of the oxygen leaves the blood and diffuses into the cells of the body for metabolic use. The elderly do not appear to have a problem with this component of the oxygen transport process.

The healthy elderly have no problem with the transport of carbon dioxide to the alveoli from the cells that generate it. Of course, those suffering from chronic obstructive pulmonary disease may accumulate carbon dioxide in their body fluids, but that is the result of inadequate alveolar ventilation, rather than a transport problem.

RENAL AND URINARY SYSTEM

The kidneys play a major role in the regulation of the composition and volume of body fluids. They also eliminate metabolic waste products and toxic substances from the body. In carrying out these functions, the kidneys form and excrete urine.

The two kidneys are located in the abdominal cavity near the back, one on the right side and the other on the left. A schematic cross section of the gross structure of one kidney is depicted in Figure 6.8. Note that each kidney has an outer cortex and an inner medulla. The kidneys are supplied with blood by large renal arteries, and each is connected to the urinary bladder by a tube called the ureter.

At a microscopic level, each kidney is comprised of about 1 million tubules called nephrons, which are the functional units of the kidneys; their structure is diagrammatically depicted in Figure 6.9. Each nephron starts in the cortex, where its closed end is in contact with a group of capillaries called a glomerulus. The nephron descends through the cortex into the medulla; it then makes a hairpin turn and returns to the cortex where, via a connecting duct, it joins the collecting duct system that ultimately joins the ureter.

Starting in young adulthood, the mass of the kidneys decreases progressively with increasing age. This loss in mass occurs primarily in the cortex of the kidney, with little loss occurring in the medulla.

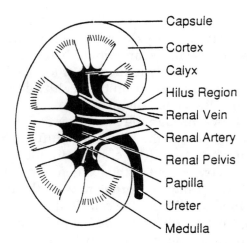

Figure 6.8 Schematic diagram of a cross section of the gross structure of the human kidney.

From "Essential Medical Physiology," (p. 338), L. R. Johnson (Ed.), 1992. New York: Raven Press. Copyright 1992 by Lippincott-Raven Publishers. Reprinted with permission.

Renal Blood Flow

To carry out the function of regulating body fluid composition and volume, it is necessary for the kidneys to process large volumes of blood (about 55% blood cells and 45% blood plasma) and, indeed, they do. Although they make up less than 0.5% of the body mass, the kidneys receive almost 25% of the cardiac output of an individual at rest. The renal arteries, which branch off the abdominal aorta, give rise to a series of smaller arteries and arterioles which, in turn, give rise to the glomeruli. (The blood pressure in the glomeruli is much higher than in most systemic capillaries; this higher blood pressure causes a fraction of the blood plasma to be filtered into the lumen of the nephron.) The blood leaves the glomeruli by way of arterioles, which then give rise to another group of capillaries (called the peritubular capillaries) that supply the nephron tubules. The blood pressure in the peritubular capillaries is lower than in most systemic capillaries, which facilitates the reabsorption of most of the filtered plasma. The venous system returns the blood, via the renal veins, to the large veins of the systemic circuit.

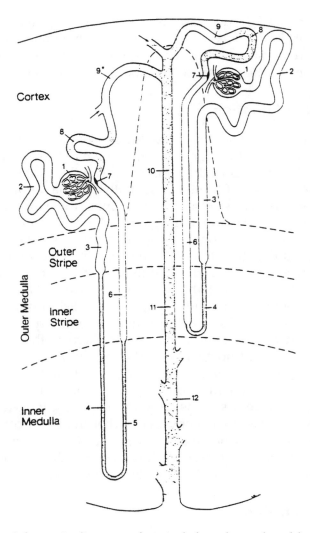

Figure 6.9 Schematic diagram of typical short-looped and long-looped nephrons together with collecting system. (not drawn to scale). 1. glomerulus and closed end of nephron; 2. proximal convoluted tubule; 3. proximal straight tubule; 4. descending thin limb; 5. ascending thin limb; 6. distal straight tubule; 7. macula densa; 8. distal convoluted tubule; 9. connecting tubule; 10. cortical collecting duct; 11. outer medullary collecting duct; 12. inner medullary collecting duct.

From "A Standard Nomenclature for Structures of the Kidney," by W. Kroz and L. Bankir, 1988. *American Journal of Physiology, 254,* p. F8. Copyright 1988 by The American Physiological Society. Reprinted with permission.

Kidney blood flow decreases with increasing age, with a decrease of more than 50% between the fourth and ninth decades of life. Much of the decrease stems from the constriction of kidney arterioles, which increases the resistance to blood flow through the kidneys.

Glomerular Filtration

The formation of urine starts with the process of glomerular filtration. About 20% of the blood plasma flowing to the kidneys is filtered across the walls of the glomerular capillaries into the lumen of the nephrons. This filtrate is essentially plasma minus its protein components. As discussed above, the driving force for this filtration is the high blood pressure in the glomerular capillaries.

In several cross-sectional studies in which subjects of different ages were studied, the glomerular filtration rate showed a decrease with increasing age, with those in the age range of 75 to 84 having about 70% that of those in the age range of 25 to 34. However, in the Baltimore Longitudinal Study of Aging, in which subjects of a wide range of ages were examined repeatedly over many years, not all exhibited a decreased rate with increasing age. Rather, there was a group who showed no change in the rate with increasing age. It has been suggested, but certainly not proved, that the decrease in glomerular filtration widely observed with increasing age is due to some disease process, such as hypertension or the long-term effects of an infectious disease experienced earlier in life. Of course, a few elderly have age-associated kidney diseases that cause such a marked reduction in glomerular filtration that dialysis or a kidney transplant is required for their continued survival.

Tubular Functions

As the filtrate flows through the nephron tubules and thence through the collecting ducts to the ureters, much of the filtrate is returned to the blood flowing in the peritubular capillaries by processes collectively called tubular reabsorption. The magnitude of this reabsorption can be appreciated by considering the fact that about 120 ml per minute of filtrate is formed in the glomeruli while the volume of urine flowing into the urinary bladder ranges from 0.5 to 2 ml per minute. Many different reabsorption systems are involved in returning the water and the chemicals dissolved in the water (called solutes) to the blood in the peritubular capillaries. These reabsorption systems are controlled so that just the correct amount of a particular solute and/or water is reabsorbed, thus maintaining the appropriate amount of that substance in the body.

In addition to reabsorption, the tubules are also able to transport some substances in the opposite direction, i.e., from the plasma in the peritubular capillary blood into the fluid in the lumen of the tubules. This process is called tubular secretion and it, too, is regulated so as to maintain the appropriate amount of the material in the body.

There are so many different transport systems involved in tubular reabsorption and secretion that little attempt has been made to study most of them in regard to the influence of age on their functioning. However, there has been sufficient study to conclude that, except when strongly challenged, most of the tubular transport systems continue to function effectively with advancing age. Some relatively common challenges will be discussed below.

Regulatory Functions

How well with increasing age do the kidneys function as a component of systems regulating body fluid composition and volume? That is the important question. The answer is: In the absence of a challenge, the kidneys of the healthy elderly retain the ability to maintain normal concentrations of plasma sodium and potassium and hydrogen ions as well as a normal extracellular fluid volume. However, when challenged by unusual circumstances, this ability is impaired.

When the diet has a very low sodium content, the kidneys conserve sodium by reabsorbing almost all the filtered sodium, i.e., the excreted urine is almost free of sodium. This ability of the kidneys to conserve sodium decreases with increasing age. As a result, the body's sodium content, as well as extracellular fluid volume and blood volume, decrease more in the elderly when so challenged than in the challenged young. Although the decreased ability to conserve sodium may be due, in part, to intrinsic changes in kidney tubular function, much of it appears to be the result of an endocrine alteration. In response to a falling blood volume, the kidneys increase their rate of secretion into the blood of a hormone called renin; this leads to the generation of angiotensin II. In turn, angiotensin II promotes the secretion of aldosterone by the adrenal cortex, and aldosterone enhances the kidneys' ability to reabsorb sodium. Under comparable circumstances, the elderly have lower blood levels of renin and aldosterone than the young, and this alteration in the renin-aldosterone system is probably the main reason that the kidneys of the old have a reduced ability to conserve sodium when challenged by a diet low in sodium. Moreover, the lower blood levels of aldosterone may also explain why the elderly tend to have high plasma levels of potassium; aldosterone promotes renal tubular secretion of potassium and thus urinary excretion of potassium.

With increasing age, the kidneys are also less able to increase the excretion of sodium in the urine to meet the challenge of a high dietary intake of sodium. As a result, high sodium intake tends to increase the extracellular fluid volume to detrimental levels in the elderly. Many reasons have been given for this reduced ability to excrete sodium; probably the most important are the decreases in both renal blood flow and glomerular filtration that occur in most people with increasing age.

A lack of drinking water leads not only to a decrease in the extracellular fluid volume and blood volume, but also to an increase in the concentration of solutes in the body fluids, because of the evaporative loss of water from the lungs during breathing and from the skin. Because of its osmotic effects, the increase in solute concentration adversely affects the functioning of the cells of the body. Fortunately, there are cells in the hypothalamus that respond to an increasing concentration of solute in the body fluids by promoting the secretion into the blood of the hormone vasopressin by the posterior pituitary gland. Vasopressin enables the kidneys to increase water reabsorption and to excrete a scant volume of urine with a much higher concentration of solute than that of the body fluids. By excreting what is referred to as a concentrated urine, the kidneys postpone the damagingly high concentrations of solutes in the body fluids until, hopefully, a source of drinking water is located. The elderly respond to an increase in body fluid solute concentration by secreting more vasopressin than do the young. However, the kidneys of the old do not increase water reabsorption in response to vasopressin as effectively as do the young; i.e., with increasing age, there is a progressive loss in the ability to excrete a scant, highly concentrated urine. Thus, when deprived of water, the elderly are more likely to develop a damaging elevation of solutes in their body fluids. This problem is exacerbated by the fact that the elderly have a blunted thirst response when suffering water deprivation, making it less likely that they will seek the drinking water needed to counter the rising solute concentration of body fluids.

With increasing age, there is also a decreased ability to cope with the intake of a large volume of water. A large volume of water decreases the solute concentration of the body fluids which, if too low, is damaging. Receptors in the hypothalamus sense the falling solute concentration and reduce or stop entirely the secretion of vasopressin. The reabsorption of water by the kidneys is thus decreased and, as a result, the kidneys excrete large volumes of urine that has a low concentration of solutes and, by so doing, prevent the solute concentration of the body fluids from falling to damagingly low levels. The ability of the kidneys to excrete a high volume of dilute urine in response to a high intake of water decreases with increasing age. Fortunately, this physiological deficit rarely causes serious problems for the elderly unless there is a coexisting disease.

The body continuously produces hydrogen ions (primarily from the metabolism of protein) and consumes them (primarily by the metabolism of components in fruits and vegetables). Hydrogen ions are highly reactive substances which, for good health, must be maintained at a nearly constant concentration in the body fluids. Acids generate hydrogen ions while bases consume them and, for this reason, the maintenance of a nearly constant hydrogen ion concentration is referred to as acid-base balance. In healthy people, the hydrogen ion concentration of the blood plasma ranges from 36 to 43 nanomoles per liter (pH 7.45–7.35). The maintenance of hydrogen ion concentration involves several processes: the participation of chemical buffers present as solutes in body fluids; the ability of the respiratory system to adjust the carbon dioxide concentration in the body fluids (which, in effect, is adjusting the concentration of carbonic acid); and the ability of the kidneys to excrete acid or base in the urine.

Based on a cross-sectional survey involving a very large number of subjects, it appears that hydrogen ion concentration in the blood increases progressively with increasing age in most healthy people; on average, it is 6 to 7% higher in 80-year-olds than in 20-year-olds. Further analysis of these data reveals that a change in kidney function is responsible for the increase. Since hydrogen ions can dissolve bone, the progressive increase in hydrogen ion concentration could be a factor in the age-associated loss of bone mass.

The ability of a person to respond to the intake of a large acid load can be tested by the oral administration of ammonium chloride. Such tests have shown that kidneys of young subjects excrete the hydrogen ions generated from the metabolism of ammonium chloride much more rapidly than do the kidneys of old subjects; thus, following ingestion of ammonium chloride, the hydrogen ion concentration in the body fluids remains elevated for a much longer period in the old than in the young. The reduced ability of the kidneys to excrete hydrogen ions both predisposes the elderly to metabolic acidosis and delays their ability to recover from it.

Urination

The urine generated by the kidneys is carried to the urinary bladder by the ureters. The urinary bladder is a smooth muscle storage bag for urine, and its structure is schematically presented in Figure 6.10. As urine flows in, the smooth muscle of the bladder relaxes to accommodate 450 to 500 ml of urine. The urine does not leak from the bladder because of two sphincters, the smooth muscle internal sphincter and the skeletal muscle external sphincter. The bladder is innervated by the sympathetic nervous system, which causes the smooth muscle of the bladder to relax and smooth muscle of the internal sphincter to contract, thus promoting the storage of urine in the bladder. The parasympathetic nervous system causes the bladder smooth muscle to contract, thus promoting the emptying of the bladder. The external sphincter is under volitional control since it is innervated by motor nerves in the same fashion as all other skeletal muscles.

The act of urination involves reflex responses to the stretching of the bladder by urine, such responses causing the bladder to contract and the internal sphincter to relax; when this is accompanied by the voluntary relaxation of the external sphincter, urine is expelled via the urethra (the tube connecting the urinary bladder to the external environment). Thus, the cerebral cortex, brain stem, and spinal cord are all involved in the process of urination.

In healthy people, aging does not adversely influence urination in a serious way, although there is a decrease in the ability to postpone voiding. Therefore, if an elderly person suffers from a serious urination problem, the cause involves a factor other than aging.

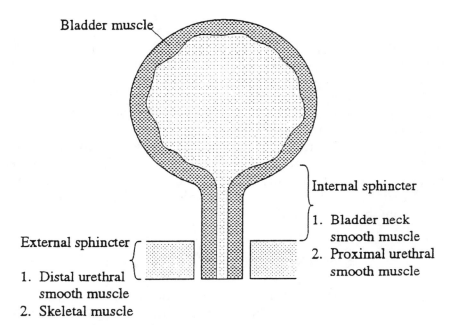

Figure 6.10 Schematic diagram of urinary bladder.

From "Geriatric Medicine," (2nd ed., p. 332), by D. W. Jahnigan and R. W. Schier, 1996, Cambridge, MA: Blackwell Science. Copyright 1996 by Blackwell Science, Inc. Reprinted with permission.

Urinary Incontinence

The involuntary passing of urine is referred to as urinary incontinence, and it is a common medical problem in the elderly. It is estimated that 15 to 30% of the community-dwelling elderly suffer from it, and, of course, the prevalence is much higher in those in nursing homes. There are two categories of urinary incontinence: transient and established. Transient incontinence has a variety of causes, the most common of which are urinary infections, drugs used for the treatment of other disorders, restricted mobility, confusional states, psychological disorders, endocrine disorders, and stool impaction. If the causative problem can be corrected, transient incontinence will disappear.

Established incontinence results from failure to empty the bladder or reduced ability of the bladder to store urine. In men, such incontinence is often due to prostatic enlargement; and in women, to diabetic neuropathy.

Prostatic Obstruction

Benign prostatic hyperplasia (BPH) commences at about age 45 and progresses with increasing age; the enlargement of the prostate gradually compresses the urethra, causing obstructed urinary flow. As a result of this obstruction, a man can suffer from hesitancy in initiating urinary flow, a weakened urinary stream, the inability to terminate urination abruptly without dribbling, incomplete emptying of the bladder, urinary frequency and/or urgency and, as mentioned above, incontinence. Indeed, it may ultimately lead to an inability to void at all, which can be fatally destructive to the kidneys, and thus the individual, unless appropriate medical intervention is obtained without delay.

GASTROINTESTINAL SYSTEM

The primary function of the gastrointestinal system is to process food so that nutrients can be absorbed into the blood and lymphatic system for distribution to the cells of the body. It also serves as a minor pathway for excretion. The system is schematically illustrated in Figure 6.11; anatomically, it is essentially a tube, starting at the mouth and ending at the anus, and the salivary glands, liver, and pancreas are connected by ducts to the lumen of this tube. To carry out its primary purpose, the system requires motor activity, glandular secretions, and digestive processes as well as absorptive processes.

Motor Activity

Both skeletal muscle and smooth muscle are involved in the motor functions of the gastrointestinal system. Skeletal muscle is involved in mastication (the chewing of food) and the initial components of swallowing, as well as components of defecation; the other motor functions are carried out by smooth muscle. Motor activities must be regulated so as to optimize digestive and absorptive processes and to coordinate with secretory processes. The sympathetic and parasympathetic nervous systems are intimately involved in this regulation.

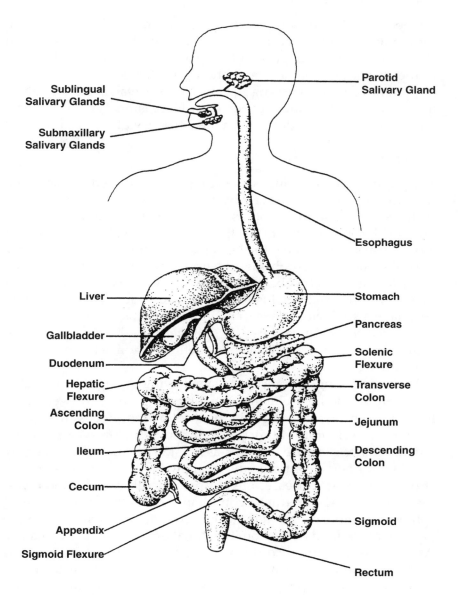

Figure 6.11 Schematic diagram of the gastrointestinal tract.
Reprinted with permission of Solvay Pharmaceuticals, Inc.

Mastication

By mechanically breaking up the food and mixing it with saliva to form a moist, lubricated bolus (a small round mass), mastication starts the digestive process and prepares the food for swallowing. In the absence of dental interventions, mastication is often greatly compromised with increasing age. Also, the skeletal muscles involved in mastication become weaker with increasing age, and consequently there is some decrease in the efficiency of mastication.

Swallowing

The process of swallowing is exceedingly complicated. It starts with the food bolus being moved by the tongue into the throat, and further movement of the bolus follows due to sequential contractions of skeletal muscles of the throat. Then relaxation of the upper esophageal sphincter enables the bolus to enter the esophagus. Once the bolus enters the esophagus, the upper esophageal sphincter contracts to prevent reflux back to the throat, and a peristaltic contraction wave pushes the bolus towards the stomach. Then, the lower esophageal sphincter relaxes, enabling the bolus to enter the stomach. Once the bolus enters the stomach, the lower esophageal sphincter closes, thereby preventing reflux of stomach contents into the esophagus.

With something so complex as swallowing, it might be expected that there would be problems with increasing age. However, in the healthy elderly, any changes in the swallowing process that occur do not cause significant functional difficulties. However, swallowing difficulties can arise with age-associated diseases that adversely affect motor nerve control of the swallowing process; these include stroke, Parkinson's disease, amyotrophic lateral sclerosis, and myasthenia gravis. Also, reduced compliance of the upper esophageal sphincter is more common in the elderly, which interferes with the passage of the food bolus from the throat down into the esophagus. Another swallowing disorder, called achalasia, relates to reduced esophageal peristaltic wave production when swallowing and a failure of the lower esophageal sphincter to open; as a result, the food bolus tends to remain lodged in the esophagus. The prevalence of this disorder increases with increasing age.

Heartburn is the result of the reflux of stomach contents through the lower esophageal sphincter up into the esophagus. It increases with age, causing elderly people considerable distress and sometimes serious complications.

Gastric and Intestinal Motility

The smooth muscle of the stomach has three principal functions: relaxation to accommodate the materials arriving from the esophagus; contractions to mix the ingested material with the gastric juice and to reduce the size of the particles of ingested material to form chyme (a gruel-like material); and contractions to propel the chyme from the stomach to the duodenum (the first part of the small intestine). The last of the three functions is referred to as gastric emptying, and this function is the most studied in regard to aging. It has long been believed that gastric emptying is slowed with advancing age. However, using advanced technologies, recent studies indicate that there is a significant age-associated change in gastric emptying only when a meal is very large. Motor activity of the small intestine (duodenum, jejunum, and ileum) appears to change little with advancing age.

Colon Motility and Defecation

Each day about 600 ml (2 pints) of chyme leave the small intestine and enter the colon (also called the large intestine). The major motor activity in the first part of the colon (the cecum and ascending colon, see Figure 6.11) involves contractions that mix the contents in the lumen of the colon and thereby aid in absorbing water from the chyme. By the time the chyme reaches the transverse colon, it has a semisolid consistency. Approximately one to three times a day, there is a mass movement of material along the colon to a region closer to the anus. This movement often coincides with the entry of food into the stomach and is called a gastrocolic reflex. The evidence indicates that in healthy people there is no change in the motor functions of the colon with advancing age.

The smooth muscle of the rectum near the anus forms a sphincter called the internal anal sphincter, which is usually in a contracted state and thus assists in preventing passage of fecal material through the anus. (The feces are composed of dietary materials that have not been absorbed, such as fiber; cells that are shed into the lumen from the wall of the gastrointestinal tract; excreted waste products; and bacteria comprising the flora of the colon.) When fecal material is forced into the rectum by the motor activity of the colon, the smooth muscle of the rectum contracts and that of the internal anal sphincter relaxes. However, defecation will not occur if the external anal sphincter (composed of skeletal muscle and thus under volitional control) does not relax. Defecation involves not only contraction of the rectum and relaxation of the internal anal sphincter, but also requires simultaneous, voluntary relaxation of the external anal sphincter and voluntary contraction of skeletal muscles of the thorax and abdomen.

The elderly frequently complain of constipation, but studies utilizing objective measures of constipation indicate that it does not occur more frequently in the elderly than in the young. This apparent discrepancy may relate to how the individual defines constipation. The elderly also frequently complain of diarrhea, but the healthy elderly do not suffer from diarrhea. When diarrhea poses a serious problem for the elderly, it is related to some disease. Fecal incontinence (the involuntary passage of feces or gas through the anus) can be a problem for the elderly. They are more prone than the young to this problem because of both higher rectal pressure, when the rectum is distended by a fecal mass, and reduced force of the anal sphincters.

Secretion

About 7 liters (7 quarts) of fluid are secreted daily by various glandular cells into the lumen of the gastrointestinal tract. These secretions contain many different chemicals whose functions are important in the digestion and absorption of nutrients.

There are three major pairs of salivary glands (parotid, sublingual, and submaxillary glands; see Figure 6.11), and they secrete about 1 liter of saliva per day through ducts into the mouth. Saliva contains amylase, which is involved in the digestion of starch and related substances; other chemicals in saliva lubricate the food bolus and protect the structures of the mouth. Much research has been done on the functioning of the salivary glands. While there is disagreement as to whether the secretion of saliva decreases with increasing age in healthy people, there is agreement that medications used to treat age-associated diseases do cause alterations in salivary secretion.

The lining of the stomach (the gastric mucosa) has various kinds of cells that collectively secrete about 2 liters of fluid per day into the lumen of the stomach. This gastric secretion contains chemicals for digestion, absorption and protection: Pepsin is involved in protein digestion; mucus has a protective function; hydrochloric acid has digestive and protective functions; and a protein called intrinsic factor is involved in the absorption of Vitamin B_{12} by the small intestine. Although it was long believed that hydrochloric acid secretion declines with increasing age, recent research has shown this is not true for the healthy elderly. The long-held, erroneous belief relates to atrophic gastritis, which does increase in prevalence and severity with increasing age. This inflammatory disease, probably caused by an autoimmune mechanism, leads to destruction of the parietal cells, which secrete hydrochloric acid; thus, those suffering this disease may secrete little or no hydrochloric acid. The parietal cells also secrete intrinsic factor; thus those who have lost most of the parietal cells will not be able to effectively absorb vitamin B_{12} and, because of this, will suffer from pernicious anemia. However, if provided with massive quantities of vitamin B_{12} as a dietary supplement, those without intrinsic factor can absorb enough of the vitamin to prevent pernicious anemia.

The pancreas secretes about 2 liters of pancreatic juice per day through a duct into the lumen of the small intestine. This fluid contains enzymes involved in the digestion of proteins, carbohydrates, and fats; it also contains bicarbonate, thus helping to neutralize the hydrochloric acid that enters the small intestine from the stomach. Although some changes in the secretion of pancreatic juice have been found in studies with healthy elderly, such changes are not great enough to affect the physiological functions of this secretion.

The liver continuously secretes bile which flows through a duct to the gallbladder, where it is stored. Within about 30 minutes of eating a meal, the gallbladder contracts and expels its contents through a duct into the lumen of the intestine. Cholecystokinin (CCK), a hormone secreted by the duodenum into the blood in response to a meal, is responsible for the contraction of the gallbladder. Micelles, which contain bile salts, phospholipids, and cholesterol, are a most important component of bile; they play a key role in fat digestion and in rendering the products of fat digestion water-soluble. The bile also contains bile pigments, heavy metals, and some drugs (in those being treated with the drugs) and serves as a pathway for the elimination of these substances.

The major effect of aging on the biliary system is the increased likelihood of gallstones. The reasons for this appear to be the change in the composition of the bile with increasing age, particularly the increase in cholesterol concentration, and the reduced contractile response of the gallbladder to CCK.

Digestion

It is necessary for most of the ingested nutrients to be converted to smaller molecules in order to be absorbed into the blood. Digestion refers to the processes that carry this out. For example, starch, a large polymer of glucose units, must be converted into glucose, a small sugar molecule referred to as a monosaccharide. Also, protein must be converted to component amino acids and small peptides containing two or three amino acids. Fats, too, must be broken down to smaller molecules and, in addition, solubilized by micelles, as discussed above. These digestive processes require enzymes that are present either dissolved in the fluids of the lumen of the gastrointestinal tract or bound to the luminal surface of the tract.

The digestion of starch starts in the mouth with the enzymatic action of salivary amylase; it ceases in the stomach when the salivary amylase is exposed to hydrochloric acid; it begins again in the small intestine where pancreatic amylase is involved. Enzymes bound to the luminal surface of the intestinal wall complete the conversion of the products of starch digestion into glucose. There appears to be no deterioration with age in the ability to digest starch.

The conversion of sucrose (table sugar) and lactose (milk sugar) to their component monosaccharides is catalyzed by enzymes (sucrase and lactase) bound to the luminal surface of the intestinal wall. There is no age-associated problem in the digestion of sucrose. In those susceptible to lactase deficiency, there is decreased ability to digest lactose with increasing age (referred to as lactose intolerance), but those not afflicted with this genetic characteristic show no age-associated difficulties in digesting lactose.

The healthy elderly have little difficulty in digesting protein and fat. However, studies have shown that the elderly do not handle massive intakes of either substance as well as do the young.

Absorption

Most of the food and water ingested and the solutes and water secreted by glands into the lumen of the gastrointestinal tract must be absorbed into the blood. Failure to do so would cause massive diarrhea, and the consequent loss of body fluid would quickly bring about the collapse of the cardiovascular system. The small intestine is the major site of absorption of most substances, although a small but significant amount of water and salts is absorbed by the colon. The cells on the luminal surface of the small intestine have many different specialized transport processes that carry out this absorption.

Tests to determine the effect of age on the capacity of the intestine to absorb glucose showed an age-associated decrease. However, the capacity to absorb glucose is so much greater than needed when eating any of a wide range of usual diets that this change does not pose a problem for the healthy elderly. This also appears to be true in regard to the absorption of the products of protein and fat digestion.

The ability to absorb calcium deteriorates at advanced ages, which causes little problem when dietary calcium intake is abundant, but presents a substantial one when intake is low. This may relate to an age-associated change in the vitamin D system. A hormone generated by the kidneys from vitamin D plays an important role in the absorption of calcium by the small intestine. There tends to be an age-associated decrease in both the amount of available vitamin D and in the ability of the kidneys to convert it to the needed hormone. Also with increasing age, there is a blunting of the action of this hormone on the calcium absorption system of the small intestine.

There is no evidence that age, in the absence of disease, causes inadequate absorption of either vitamins or minerals other than calcium. However, it must be noted that there have been few well-conducted studies on the influence of age on the absorption of these substances.

ENDOCRINE AND METABOLIC FUNCTION

The endocrine glands secrete many different hormones to regulate the physiological activities of the body. The regulation of body fluid volume and composition by aldosterone and vasopressin has already been discussed, as has the regulation of gallbladder contraction by cholecystokinin (CCK). Since space does not permit coverage of all the hormones and the many processes they regulate, this discussion will concentrate on those that are likely to be involved in aging. Many of the age-associated alterations in endocrine function relate to metabolic processes, i.e., the complex of chemical events that underlie living processes. Therefore, much of this discussion will focus on metabolism, both *per se* and its regulation by the endocrine system.

Energy Metabolism

It was pointed out in Chapter 4 that almost all living organisms require a continuous supply of ATP as an energy source for physiological activities. A general outline showed how this need is met primarily by the oxidative metabolism of fuels (i.e., carbohydrates and fats as well as proteins) from dietary sources and body stores. Energy metabolism refers to the collective use of all these fuels, while metabolic rate refers to the rate of energy expenditure (i.e., the rate of fuel use).

Total daily energy expenditure (i.e., daily metabolic rate) decreases with increasing adult age. In a thermally neutral environment (i.e., one in which environmental temperature does not influence metabolic rate), total daily energy expenditure is comprised of three major components. These include: the basal metabolic rate (BMR); fuel use due to physical activity; and diet-induced thermogenesis (i.e., the increase in metabolic rate due to processing of food). Of course, in a cold environment, shivering thermogenesis becomes a significant component of the daily energy expenditure.

Conceptually, the basal metabolic rate refers to the energy expenditure needed to carry out the basic functions of living; it is measured by determining the rate of energy expenditure in a resting (not sleeping) person fasted for 12 to 14 hours in a thermally neutral environment. It usually accounts for about 60 to 75% of the daily energy expenditure. The BMR decreases with increasing age. It was felt that this decrease might be due to a decrease in thyroid gland function with increasing age, because hypothyroidism is known to decrease the BMR. The level of thyroid hormones in the blood has been the subject of much study and although there is not full agreement on the findings, it seems clear that the decrease in BMR with increasing age probably does not stem from a decrease in thyroid hormone availability. Indeed, if the BMR is expressed per kilogram of lean body mass, there is not a marked decrease with age in the BMR. Thus, it seems likely that the major reason for the decreased BMR is the change in body composition with age, namely the decreasing skeletal muscle mass and the increasing adipose tissue mass, the latter having a much lower metabolic rate than most other tissues.

The contribution of physical activity to the metabolic rate is significant, but variable, ranging from an estimated 15 to 30% of the total daily energy expenditure. This includes not only vigorous exercise but also the usual motor activities of daily living as well as the increase in metabolic rate that continues for some time after physical activity ceases. The effect of age on this component of energy expenditure has not been measured directly; however, it is believed to decrease significantly, based on extensive evidence that physical activity decreases with age.

Diet-induced thermogenesis refers to the fact that the metabolic rate is increased by the ingestion of food. This increase accounts for about 7 to 13% of the total daily energy expenditure. While aging does not appear to cause a major change in diet-induced thermogenesis, this subject remains to be carefully studied.

Carbohydrate and Fat Metabolism

Glucose is the major monosaccharide formed during digestion of carbohydrates. Moreover, much of the other two monosaccharides (fructose and galactose) generated during carbohydrate digestion is converted to glucose during absorption by the small intestine or while passing through the liver on the way to the systemic circulation. Thus, glucose is by far the predominant carbohydrate fuel used by the cells of the body.

The metabolism of glucose is usually evaluated by the oral glucose tolerance test. In this test, a standard amount of glucose is ingested by the subject, and changes in the plasma concentration of glucose are measured over the next 2 hours. Older people usually show higher plasma glucose levels during this test than do young people; i.e., they have less ability to utilize glucose. In medical circles, this is referred to as impaired glucose tolerance.

Insulin, which is secreted by endocrine cells in the pancreas, promotes the body's use of glucose as fuel and its storage as glycogen in liver and muscle. It was suspected that the decrease in ability of the elderly to utilize glucose might be due to a decreased ability of these endocrine cells to secrete insulin. However, careful study has shown that the age-associated impairment of glucose tolerance is not due to a reduced ability to secrete insulin. Rather, there is an age-associated decrease in the response of many cells of the body to insulin, and this is the cause of the impaired glucose tolerance. This loss in responsiveness to insulin is called increased insulin resistance.

Although an age-associated increase in insulin resistance is a common occurrence, age *per se* is not the major reason for it. The major causal factors appear to be the decrease in physical activity and the increase in adipose tissue mass that are usually associated with advancing age. Indeed, there is little increase in insulin resistance in the elderly who are physically fit and relatively lean.

Fat, which biochemists call triglyceride, is the other major fuel of the body. It appears that the ability to use fat as fuel is not altered with increasing age. However, fat metabolism is associated with the metabolism of cholesterol, and it is this association that has relevance to aging.

Both fat and cholesterol are almost insoluble in aqueous solutions such as the body fluids. In order to be transported in the plasma of the blood, they are "packaged," together with particular proteins called apolipoproteins, to form lipoprotein molecules that are soluble in plasma and other body fluids. Elevated levels of LDL (low density lipoproteins) increase the risk of atherosclerotic diseases, such as coronary heart disease and stroke. On the other hand, high levels of HDL (high density lipoproteins) lower the risk of these diseases. There is an age-associated increase in the plasma concentration of LDL in most men until about age 50, and in women until about age 70. Women have an age-associated increase in plasma HDL concentration until about age 60, while men have lower levels of HDL than women throughout adult life; however, men show an increase in HDL from age 50 to 65, which narrows the gender difference.

Although most elderly people exhibit the age changes in carbohydrate and fat metabolism just described, a small fraction do not. Those who are physically fit and have only a small increase in body fat do not undergo the age changes in plasma lipoproteins that increase the risk of atherosclerotic disease nor, as stated above, do they exhibit an appreciable increase in insulin resistance.

Unfortunately, a large subset are not physically fit and show marked changes in carbohydrate and fat metabolism. Many suffer from a collection of signs and symptoms referred to as Syndrome X: These include high blood sugar levels, high blood insulin levels, insulin resistance, high blood LDL and low blood HDL levels, high blood pressure, and high body fat content distributed preferentially in the abdominal region (a potbelly). Clearly, these people are at risk for coronary heart disease and stroke. Indeed, it is felt that the abdominal obesity may play a causal role in Syndrome X. Fortunately, the signs and symptoms of this syndrome can be markedly attenuated by a reduction in body fat and an increase in physical activity.

Type II diabetes (non-insulin-dependent diabetes mellitus), which has many of the characteristics of Syndrome X but in an exaggerated form, increases in prevalence with increasing age. It is estimated that 20% of individuals in the age range of 65 to 74 suffer from Type II diabetes. It is logical to think that this type of diabetes is an outgrowth of Syndrome X, but clear evidence in support of this hypothesis is lacking. In addition to a marked insulin resistance, there is also an impaired ability of the pancreas to secrete insulin in those with Type II diabetes.

Protein Metabolism

With advancing age, there is a decrease in the rate of protein synthesis and protein degradation; i.e., the rate of protein turnover decreases. This poses a problem because it increases the length of time a protein molecule spends in the body. As they reside in the body, protein molecules are gradually damaged by oxidation, glycation, heat, and other factors. Thus by increasing the average length of time a protein molecule spends in the body, the age-associated decrease in the rate of protein turnover acts to increase the amount of damaged protein molecules. It also seems likely that the age-associated decrease in muscle mass relates, at least in part, to the decrease in protein synthesis.

Much of the age-associated decrease in protein synthesis may be due to the decreased secretion of growth hormone by the anterior pituitary gland. Growth hormone does not directly influence protein synthesis, but rather it promotes the secretion of an insulin-like growth factor, IGF-I, by the liver and other tissues. It is IGF-I that promotes protein synthesis. IGF-I levels decrease with increasing age in parallel with the decrease in growth hormone.

Growth hormone also promotes the use of fat as fuel, thereby sparing the use of protein for fuel. Thus the low levels of growth hormone may decrease the use of fat relative to protein, thereby contributing to the increase in body fat and the decrease in muscle mass that occur with advancing age.

Stress

Harmful physical events, or those perceived as harmful, cause stress in humans and other organisms. A hallmark of aging is the decreased ability to successfully cope with stress.

This may well stem from the age-associated decrease in the expression of stress response genes, in particular, the genes expressing heat shock proteins. The heat shock proteins are produced in response to a spectrum of stressors, including heat, and they protect cells from the potential damage of many different extrinsic and intrinsic factors. It should be noted that they serve this purpose in almost all living organisms.

In addition to this basic cellular mechanism, there are systemic responses involving the nervous system or the endocrine system, or both, that may help animals to cope with stress. Although these systemic responses are protective, they can also be damaging in their own right, if their response is too intense or prolonged. Indeed, some feel that these protective mechanisms may accelerate the rate of aging.

The sympathetic nervous system, including the adrenal medulla, plays an important role in the body's response to stress. Its activity increases under conditions of stress, and it enhances cardiovascular function as well as mobilizing fat and carbohydrate fuel from body stores, actions that aid in coping with stress. With increasing age, there is increased activity of the sympathetic nervous-adrenal medullary system in response to stress. However, at least at some target sites, the response of the cells to this system is blunted with increasing age (e.g., the response of the heart).

In addition, the glucocorticoid response is critical in the individual's ability to successfully cope with stress. The hypothalamus reacts to stress by sending a hormone directly to the anterior pituitary gland, causing it to secrete ACTH (adrenocorticotrophic hormone). ACTH promotes the secretion of cortisol (a glucocorticoid) by the adrenal cortex. It is believed that glucocorticoids assist in coping with stress, at least in part, by protecting against the damaging action of other defenses, such as the deleterious aspects of immune and inflammatory processes. On the other hand, it is known that excessive levels of glucocorticoids are damaging in their own right and can promote aging. There is little or no increase in the level of plasma cortisol with advancing age in nonstressed, healthy people. Under conditions of stress, plasma cortisol levels increase more in the elderly than in the young, and the increase is more prolonged. However, whether this increased cortisol response is beneficial or detrimental remains to be determined.

DHEA (Dehydroepiandrosterone) and Melatonin

Studies have shown that there is a marked decrease in plasma concentration of DHEA and melatonin with advancing age. Based on these findings, it has been claimed that a decrease in the concentration of one or both of these hormones plays a major role in aging.

DHEA is secreted by the adrenal cortex with the rate of secretion regulated by ACTH. Tissues in the body convert DHEA to estrogen (female sex hormone) and testosterone (male sex hormone), but the extent of these conversions seems to vary greatly among individuals. The blood level of DHEA peaks at about age 20 and decreases thereafter, so by age 60 the level is 60 to 70% of that at age 20. Low levels of blood DHEA have been associated with some forms of cancer, cardiovascular disease, dementia, diabetes, obesity, and osteoporosis. These associations and the effects noted when DHEA is pharmacologically administered to animals and humans are the basis for the claim that low levels of DHEA have a causal role in aging, and that its pharmacological use can slow aging. This latter possibility will be discussed further in Chapter 7.

Melatonin is a hormone secreted by the pineal gland, which is part of the forebrain (see Figure 6.3). It is a factor in the regulation of circadian rhythms (daily patterns of activity) in most species, probably including humans. It is believed to serve in inducing sleep, and the low levels associated with aging could be a factor in insomnia in the elderly. Melatonin is secreted during the night (i.e., when the individual is in the dark), and both the amount secreted and the duration of its secretion have been found to diminish with increasing age. Indeed, the decreased secretion of melatonin is claimed by some to be a major cause of aging, and a variety of reasons have been cited for this. The most convincing argument is that melatonin is a powerful antioxidant and that falling levels of melatonin result in an increase in oxidative damage. The pharmacological use of melatonin has been heralded (or touted, depending on one's belief in its efficacy) as an antiaging agent. Indeed, it is currently being used for this purpose. This will also be discussed further in Chapter 7.

FEMALE REPRODUCTIVE SYSTEM

The ovaries are the primary female sex organ. Each of the two ovaries is about the size of a walnut, and they are located in the abdominal cavity. At puberty, they contain about 400,000–800,000 primordial follicles, each comprised of a cell, which may mature into an ovum, surrounded by a layer of granulosa cells. Secondary sex organs include, among others, the oviducts (along which an ovum travels from the ovary to the uterus); the uterus (in which an embryo and then fetus can develop), the vagina (the receptive organ for sexual intercourse and the birth canal), and the breasts.

Beginning at puberty and ending at menopause, women exhibit a continuously repeated (unless interrupted by pregnancy) monthly reproductive cycle called the menstrual cycle. This complex cycle involves the action and interplay of at least five hormones: estrogen (specifically, estradiol) and progesterone; follicle stimulating hormone (FSH) and luteinizing hormone (LH); and gonadotrophin releasing hormone (Gn-RH). The first day of menstruation is considered the beginning of the menstrual cycle. Menstruation, which refers to the sloughing of part of the endometrium (the layer of cells lining the cavity of the uterus) along with blood through the vagina, lasts about 4 days. During menstruation, several follicles begin to develop, but only one will become dominant and, at midcycle, release a mature ovum into the oviduct, a process called ovulation; the others will undergo degeneration. During the first half of the menstrual cycle, increasing levels of FSH promote the secretion of estrogen. Rising estrogen levels, in turn, cause secretion of LH which triggers ovulation about 14 days after the start of menstruation. The follicle that yielded the ovum is then transformed into a corpus luteum which secretes estrogen and progesterone, thereby preparing the uterine endometrium for possible implantation of a fertilized ovum. If the ovum is not fertilized by male sperm, the corpus luteum degenerates in about 2 weeks, thereby eliminating this source of estrogen and progesterone, and resulting in menstruation.

Fertility

The capacity of a woman to conceive and give birth is referred to as her fertility. In studying the influence of age on fertility, what is usually measured is the number of live births per thousand women in a particular age range. However, because of confounding factors such as contraception and abortion, this measurement may not be an accurate index. Communities of Hutterites (a religious sect that originated in Moravia and now living in Canada and the northwestern United States), whose beliefs prohibit contraception and abortion, provide a population in which some of the confounding factors are circumvented. In these communities, the live birth rate per thousand women is highest in the age range of 25 to 34, which indicates a drop in fertility at ages 35 and greater. This conclusion is strengthened by studies of women in programs of artificial insemination, where the fertility of the male partner and the frequency of coitus are not confounding variables. The fact that women over age 50 rarely give birth provides clear evidence that there is an almost complete loss of fertility by that age. At ages over 30, there is an increase in spontaneous abortions, and about 50% of the aborted embryos and fetuses show genetic abnormalities. Moreover, with increasing age of the mother, starting at about 30, there is a progressive increase in the number of newborns with chromosomal abnormalities resulting in diseases such as Down Syndrome.

Menopause

Biologists define the menopause as the natural permanent cessation of periods of menstruation. It is the signpost that the menstrual cycle is no longer functioning and that ovulation has ceased. In general lay usage, however, menopause is a synonym for "change of life," and both terms refer to the years immediately before and after the cessation of menstruation, during which a variety of symptoms are experienced. On the other hand, biologists refer to those years as the premenopausal period and the immediate postmenopausal period, respectively. The biologists' definitions will be used throughout this discussion.

Age of Occurrence

It is difficult to get reliable information on the age at which menopause occurs. In the 2 to 8 years preceding the menopause, the typical

28-day menstrual cycle becomes variable, ranging from less than 28 days to more than 60 days. Based on studies that have been appropriately designed and well executed, the median age of menopause appears to be in the tight range of 49 to 50 years of age for populations of all races. Reports of much younger ages of menopause (for populations of Punjabis in India, Melanesian women in New Guinea, and Indian women in Mexico) may relate to health or nutrition rather than to a racial difference. Indeed, there is strong evidence that malnutrition results in the early onset of menopause. Women who have smoked for much of their adult life undergo menopause about 1 year earlier than nonsmokers.

Endocrine Changes

In the 2 to 8 years preceding the menopause, plasma estradiol levels are lower than at younger ages, and FSH levels are elevated. Menopause is believed to result primarily from the ovary's loss of ability to secrete estradiol. Following menopause, plasma estradiol levels are low, while plasma FSH and LH levels rise markedly; a year after menopause, FSH levels are 10 to 15 times higher than in young women, and LH levels are three times higher. These high levels of FSH and LH appear to be the result of an increased frequency and magnitude of pulses of GnRH secretion by the hypothalamus, which is no longer regulated by the estradiol secreted by the ovary.

Changes in Secondary Sex Organs

Following the menopause, pubic hair decreases and the vagina shortens, loses elasticity, and is at increased risk of bacterial infections and mechanical damage. The oviducts shorten, and their diameter decreases. There is atrophy of the glandular structure of the breasts, with replacement by adipose tissue. The uterus reduces in size, and there is atrophy of the cervix.

The incidence of breast cancer increases with advancing age. While cancer of the uterine cervix does occur in elderly women, it peaks from ages 48 to 55. Peak age range for cancer of the uterine endometrium is 60 to 64.

Symptoms

Some women have few symptoms associated with menopause, while others suffer many problems. The most common is called the "hot flush" or "hot flash," which is characterized by blushing and a sensation of heat (both due to dilation of skin blood vessels which increases blood flow to the skin) as well as inappropriate sweating. The intensity of this symptom varies among women, from an occasional, transient sensation of warmth to periodic episodes of a sensation of heat, drenching sweats, and tachycardia (rapid heart rate). In some women, these episodes result in disturbed sleep, fatigue, and irritability. The intensity of "hot flashes" peaks during the first 2 years after cessation of menstruation, but in some women the problem continues for as long as 10 years. Much effort has been made to determine the physiological basis of these episodes of altered blood flow to the skin, sweating, and rapid heart rate; but to date, there is no general agreement as to the mechanism involved.

A number of psychological symptoms are associated with the menopause, such as apprehension, apathy, depression, excitability, fear, loss of libido, rage, and bouts of uncontrollable crying. Usually these symptoms gradually disappear during the postmenopausal years.

Hormone Replacement Therapy

Many of the menopausal changes in the secondary sex organs and in other processes, such as accelerated bone loss, appear to be, at least in part, the result of estrogen deficiency. Thus it seemed likely that estrogen replacement therapy might attenuate undesirable menopausal changes, and it has been found effective in regard to certain problems; e.g., it often reduces both the frequency and duration of "hot flashes," including the associated sleep problem. Indeed, this therapy is believed to have many benefits, but also some dangers, both of which will be discussed further in Chapter 7.

MALE REPRODUCTIVE SYSTEM

The two testes are the primary male sex organ. They are located in the scrotal sac, which hangs outside the body from the pelvic region of the abdomen. The testes have two functions: the generation of sperm and the production of the major male sex hormone, testosterone. Secondary sex organs include the penis, seminal vesicles, and prostate.

Testosterone, which is secreted by the Leydig cells of the testes, is involved in sperm generation. It is also required for both the development and maintenance of the secondary sex organs. In addition, testosterone regulates other male sex characteristics, which include: distribution of body hair (including the head's hair loss pattern, often seen at a relatively young age); the pitch of the male voice; muscle mass and structure; skeletal mass and structure; distribution of body fat; and the libido. The anterior pituitary secretes two hormones that act on the testes, FSH (follicle stimulating hormone) and LH (luteinizing hormone). These hormones are also secreted by the female, and their names reflect their actions in female reproduction. In men, FSH promotes the generation of sperm by the seminiferous tubules of the testes, while LH promotes the formation of testosterone by the Leydig cells. The secretion of these hormones can be affected by a wide range of neural influences.

Age Changes in the Testes

There is atrophy of the seminiferous tubules with advancing age. In the age range of 20 to 30, 90% of the seminiferous tubules contain sperm, while in men over 80, only 10% of the tubules contain sperm. Nevertheless, some sperm are found in the ejaculate of about half the men in the age range of 80 to 90. Indeed, there is evidence that at least some men are still fertile at 80-plus years.

While there have been reports that the Leydig cells decline in number with advancing age, there are also reports that the number does not change with age. The information on the effects of age on the plasma concentration of testosterone is complicated for a variety of reasons, including marked individual variation among men of all age ranges. The mean reduction in plasma free testosterone (which is biologically active since it is not bound to plasma protein) is about 30% between the ages of 40 and 70, but some older men have levels well within the range found in young men.

Age Changes in the Secondary Sex Organs

The secretory activities of the seminal vesicles and the prostate decrease with increasing age, and thus the ability to produce semen is diminished. Although the capacity for ejaculation is maintained in most men even at very advanced ages, its characteristics change with age. The force of release of the semen from the penis is markedly reduced, as is the volume released. As discussed earlier, benign prostatic hyperplasia occurs with advancing age and causes problems with urinary function. It is rarely seen in men younger than 40, but increases markedly with further advancing age, with about 90% of men suffering the problem by age 80. Prostate cancer is rare before age 50, but is a common form of cancer with further increasing age. Although in many men it progresses slowly (particularly at advanced ages), it is the third most common cause of death in men over 55.

Impotence increases with increasing age. There is a slowing of arterial filling of the penis and an increased venous drainage from the penis, thus yielding a less firm erection. This ultimately can result in an erectile response that is inadequate for entrance into the vagina, hence impotence. By a conservative estimate, 5 to 10% of men in the sixth decade of life suffer from impotence; the figure rises to 20% in the seventh decade, 30 to 40% in the eighth decade, and 50% in the ninth decade. A prescription drug, recently available in the United States and some other countries, successfully treats this problem in many men, and other similar drugs are currently under development.

IMMUNE SYSTEM

An elaborate set of processes, collectively called the immune system, has evolved to protect higher animals from the damaging actions of bacteria, fungi, protozoa, and viruses. This system must be able to recognize such invaders and destroy them—but not attack the body's own cells. Not only can the immune system accomplish this, but in addition, it is often able to recognize and destroy cancer cells, i.e., the body's own cells that have undergone alteration, becoming mutant cells.

Two classes of lymphocytes play key roles in immune function, the B-lymphocytes, which originate in the bone marrow, and the T-lymphocytes, which mature in the thymus. Upon leaving the bone marrow and thymus, the lymphocytes are distributed to many sites in the body, such as the lymph nodes, adenoids, tonsils, and Peyer's patches in the intestinal wall. They are also found in the blood (where they comprise about 20% of white blood cells), and in the lymph (where 99% of white blood cells are lymphocytes). Other important cells in immune function are the macrophages and a major class of white blood cells called neutrophils.

The immune system goes into action upon the arrival of a specific macromolecule (called an antigen) of a foreign invader. It acts in three fundamentally different modes: humoral immunity; cellular immunity; and the secretion of lymphokines (a group of stimulatory proteins).

The B-lymphocytes are responsible for humoral immunity. Foreign invaders (e.g., bacteria) have highly specific antigen molecules to which a very small specific subset of B-lymphocytes respond. This response involves the subset's secretion of a unique protein molecule called an antibody. These antibodies bind to the antigens of the invaders, which leads to the destruction of the invaders, including their engulfment (phagocytosis) by macrophages and other cell types (e.g., neutrophils).

A subclass of T-lymphocytes called CTLs is responsible for cellular immunity. Specific CTLs recognize cells with specific antigens (e.g., cells infected with a specific virus will have antigens related to that virus), and they bring about apoptosis of the infected cells.

Another subclass of T-lymphocytes called T_H-cells are responsible for the secretion of stimulatory proteins known as lymphokines; T_H-cells respond to an antigen by secreting lymphokines that promote an increase in both the number and responsiveness of specific B- and T-lymphocytes as well as macrophages.

The immune system has memory, a phenomenon well illustrated by the B-lymphocytes. On the first encounter with a particular antigen, the responding B-lymphocytes not only secrete specific antibodies but they also undergo cell division, yielding many new B-lymphocytes capable of secreting that specific antibody. Indeed, some of the B-lymphocytes mature into what are called plasma cells, which secrete the antibody at a high rate. However, in addition to the plasma cells, memory cells are formed that persist in the blood and lymph for the rest of the person's life. These memory cells can respond rapidly to another encounter with the same antigen, thus providing a faster and more effective immune response than during the first encounter. (Memory T-lymphocytes are also formed during the first encounter with a particular antigen.)

While changes in the immune system occur with increasing age, there is great individual variation in the extent of such changes. However, it is fair to conclude there is diminished functioning of the immune system with increasing age in many, if not most, people.

Age Changes in the Lymphoid Tissues

The thymus gland, located behind the breast bone just below the neck, changes markedly with increasing age. It reaches its largest size just before the onset of puberty, and decreases in size thereafter. With increasing age, the cellular elements of the thymus are gradually replaced by adipose tissue, and the production of thymic hormones ceases by about age 40. There are no major changes in the bone marrow or the peripheral lymphoid tissues with advancing age in healthy people. However, lymphomas (tumors of the lymphoid tissues) and many leukemias exhibit an age-associated increase.

Age Changes in Lymphocytes

There appear to be two populations of T-lymphocytes: those that proliferate in response to antigens, and those that fail to respond and ultimately undergo apoptosis. Both young people and old people have these two populations of T-lymphocytes, but with increasing age, the nonresponder fraction increases. The net result is an age-associated impairment in the ability to increase the number of T-lymphocytes that can respond to a particular antigen.

The amount of antibody secreted by a given number of B-lymphocytes is known to decrease with increasing age This appears to be due to the fact that there are also two populations of B-lymphocytes, one that responds to an antigen and one that does not, and the fraction of nonresponders increases with increasing age. The extent to which the age change in B-lymphocyte function is secondary to altered T_H-cell function is not known, but it could well be substantial.

Disease and the Aging Immune System

There is little doubt that deterioration of the immune system contributes to illness in the elderly. It is a factor in the increased susceptibility of the elderly to infectious diseases, such as pneumonias, urinary tract infections, and tuberculosis. In addition to underlying the higher incidence of these infectious diseases, the deterioration of the immune system also results in an increased morbidity and mortality from such diseases. Of course, factors other than immune function are also involved in this increased vulnerability, such as age changes in the functioning of the respiratory, cardiovascular, and urinary systems.

It is reasonable to hypothesize that the deterioration of the immune system could be a factor in the increasing incidence of cancer with increasing age. Specifically, it seems likely that the deterioration of immune surveillance fails to effectively eliminate mutant cells, thereby increasing the risk of cancer. However, the validity of such a scenario has yet to be established.

The age-associated increase in autoimmune diseases, such as rheumatoid arthritis, systemic lupus erythematosus, and glomerulonephritis, certainly results from deterioration of the immune system. Specifically, there is a loss in the ability to distinguish between self and non-self.

THERMOREGULATION

A healthy person has a core body (rectal) temperature of about 98.6°F (37.0°C). Maintaining a relatively constant body temperature involves a complex control system. This begins with receptors that sense the temperature, which include cold and warm receptors in the skin as well as receptors in the hypothalamus that sense deep body temperature. All of this sensory information is integrated by the central nervous system, with the hypothalamus playing a major role. A set of responses brings about the appropriate decrease or increase in body heat. These responses include constriction of the blood vessels of the skin (which serves to conserve body heat) or dilation of these vessels (which increases the loss of body heat). Also, when the need arises, skeletal muscle contractions (referred to as shivering) are brought into play to increase the rate of heat production by these muscles. Under environmental conditions tending to cause body temperature to rise, the sweat glands are directed to secrete sweat, which cools the body by evaporative loss of water from the skin. Perceiving information coming to it from the thermal receptors, the central nervous system also evokes appropriate behavioral responses that facilitate the maintenance of body temperature, such as putting on a coat and seeking shelter, or shedding a sweater and turning on an air conditioner.

With increasing age, there is a deterioration in the ability to regulate body temperature and, thus, to meet the challenge of different thermal environments. It must be noted that the extent of this deterioration varies among individuals, depending on health, physical fitness, and lifestyle factors, such as alcohol consumption and smoking.

In a hot environment, the elderly have a reduced ability to redistribute blood flow from the core of the body to the skin, due to structural changes in the skin blood vessels and to decreased capacity to constrict the vessels supplying the viscera. In addition, the sweating response is decreased with increasing age. In a cold environment, the elderly show a decrease in constriction of skin blood vessels and a reduced shivering ability.

Elderly people have a much greater risk of developing heat stroke (hyperthermia with a rectal temperature of 106°F or higher) and hypothermia (a rectal temperature of 95°F or less) than do young people. One reason for this greater risk is the deterioration of the physiological processes involved in maintaining body temperature. However, the main reason the elderly are more vulnerable to extremes in environmental temperature is the blunting of their perception of ambient temperature. This loss is critical, because humans utilize behavioral responses rather than physiological responses as their major mode of coping with temperature extremes. The blunting of temperature perception means the elderly are less likely to make effective behavioral responses, such as seeking appropriate clothing and shelter.

ADDITIONAL READING

Jahnigen, D. W., & Schrier, R. W. (Eds.) (1996). *Geriatric medicine* (2nd ed.). Cambridge, MA: Blackwell Science.

Hazzard, W. R., Bierman, E. L., Blass, J. P., Ettinger, W. H., Jr., & Halter, J. B. (Eds.) (1994). *Principles of geriatric medicine and gerontology.* New York: McGraw-Hill.

Lamberts, S. W., van den Beld, A. W., & van der Lely, A. J. (1997). The endocrinology of aging. *Science, 273,* 419–424.

Masoro, E. J. (Ed.) (1995). *Handbook of physiology, Section 11, Aging.* New York: Oxford University Press.

Spirduso, W. W. (1995). *Physical dimensions of aging.* Champaign, IL: Human Kinetics.

Tallis, R., Fillit, H., & Brocklehurst, J. C. (Eds.) (1998). *Brocklehurst's textbook of geriatric medicine and gerontology* (5th ed.). London: Churchill-Livingstone.

Whitbourne, S. K. (1996). *The aging individual.* New York: Springer.

7

Possible Interventions to Retard Aging

The concept of a magic bullet that prevents or reverses aging has long engrossed humans. It was during his search for the "fountain of youth" that Ponce de Leon discovered Florida in 1513, and today a visit to any pharmacy or health food store shows that people are still preoccupied with the idea of a potion that can restore youth. Indeed, newspapers and TV news constantly report on new "cures" for aging. One example is the research on telomerase, an enzyme said to hold promise of slowing, if not reversing, aging. A critical assessment of currently available facts certainly does not support this optimistic view. As discussed in Chapter 4, although telomerase does prevent telomere shortening, the evidence that this shortening plays any role in organismic aging is weak to the point of nonexistence. Such unwarranted claims of "cures" for aging are unfortunate, and over the years the field of gerontology has garnered a bad name among biological scientists because so many such claims have falsely raised the hopes of the general population. Thus it is imperative that possible antiaging interventions be weighed critically.

DIETARY RESTRICTION IN RODENTS

Of the many claims for antiaging interventions, the one that has best stood the test of time is the retardation of aging in rats and mice achieved by reducing their food intake. In 1935 Clive McCay and his colleagues published the first convincing evidence that reducing food intake, starting soon after weaning, increases the length of life in rats. In the ensuing years, these findings have been greatly expanded by many other investigators using a variety of different genetic strains of rats and mice as well as hamsters. The survival curves in Figure 2.2 for populations of male F344 rats, studied in my laboratory in the 1970s, show the marked increase in length of life of rats fed 40% less food than rats fed *ad libitum* (allowed to eat as much food as they want). Gerontologists refer to this effect as the antiaging action of dietary restriction (DR). Because DR was initiated soon after weaning in the early studies, many felt that its antiaging effects were due to a retardation of growth and development. However, it was subsequently shown that DR not only extends length of life when initiated in young growing rats and mice, but is almost as effective when initiated in the young adult rat and mouse. Indeed, DR even shows a significant, though less marked, antiaging action when initiated in rodents entering middle age. Assessment of the effects of DR on the age-associated increase in age-specific death rate further attests to its antiaging action; the mortality rate doubling time is about 100 days for rats fed *ad libitum* and about 200 days for those fed 40% less food.

In addition to increasing the length of life, DR also slows the age-associated deterioration of most physiological processes in rats and mice. The following examples provide a glimpse of the breadth of the age changes that are retarded: decrease in physical activity, loss in ability to negotiate a maze, loss of brain receptors for neurotransmitters, deterioration of immune function, decrease in growth hormone secretion, alterations in gene expression, and many, many more. The vast amount of research in this area can be summarized by saying that dietary restriction maintains most physiological systems in rodents in a youthful state at advanced ages.

DR also delays the age of onset and/or the rate of progression of most age-associated diseases in genetic strains of mice and rats that are predisposed to those particular age-associated diseases. These include, among others, a variety of different cancers, kidney diseases, and autoimmune diseases.

An obvious question was whether the antiaging actions of DR were due to the reduced intake of a specific dietary component, such as fat or possibly a mineral or a specific vitamin. Extensive studies have led to the conclusion that it is not the restriction of a particular food substance that is significant, but rather the restriction of energy (calorie) intake. These findings have led some investigators to refer to this intervention as caloric restriction (CR) rather than DR.

It is to be expected that DR would decrease body fat. Studies in which body fat was measured show that both the absolute amount and the percent fat content of rats and mice are decreased by DR. Because there is evidence that excess body fat in humans is a risk factor for premature death, many felt that the reduction in body fat underlies the increase in length of life of rats and mice seen with DR. However, examination of the relationship between body fat content and length of life in dietary restricted and *ad libitum* fed mice and rats revealed that reduction in body fat content is not causally related to DR's ability to increase length of life.

Another widely held view was that DR retards senescence by decreasing the metabolic rate. As discussed in Chapter 4, the concept that the rate of aging inversely relates to the metabolic rate was first proposed by Rubner in 1908 and it has had advocates ever since. Moreover, it is known that when a person reduces food intake, the metabolic rate per square meter of body surface decreases. While the metabolic rate of rats per gram of lean body mass has been found to decrease upon initiation of DR, the decrease is transient; and within 2 weeks, there is no difference in metabolic rate between rats on restricted food intake and those eating *ad libitum*. Since it has been shown that sustained reduction of food intake over much of the life span is needed to significantly increase the length of life in rodents, a decreased metabolic rate is obviously not the factor responsible for the increase in longevity caused by DR. However, the idea of decreased metabolic rate underlying DR's antiaging action is so conceptually attractive that some gerontologists still embrace it.

Indeed, there is evidence that DR alters some of the characteristics of fuel use (other than metabolic rate) and that one or more of these alterations may underlie its antiaging action in rodents. For most of the life span, the daily plasma glucose level of the DR rats is about 15% below that of the *ad libitum* fed rats. Moreover, DR causes about a 50% decrease in plasma insulin levels. Surprisingly, despite the decrease in plasma glucose and insulin levels, DR does not cause a decrease in the rate of glucose use as fuel per gram of lean body mass. In humans, abnormally high levels of plasma glucose cause damage, such as basement membrane thickening and impaired immunity, which is similar to what occurs during aging; high insulin levels also cause such aging-like damage as atherosclerosis and cancers. Thus the ability of DR to enable effective carbohydrate fuel use at reduced levels of plasma glucose and insulin may well play an important role in its antiaging action in rodents.

In addition, DR in rodents decreases the age-associated increase in oxidative damage resulting from the use of oxygen in the metabolism of fuels; it retards the cellular accumulation of lipid peroxides, the increase in the carbonyl content of proteins, and the increase in oxidatively damaged DNA. DR protects the rodent in this regard by decreasing the generation of reactive oxygen molecules and increasing the ability to detoxify such molecules.

Insights on how DR retards aging have also been sought by considering the possible basis of the evolution of its antiaging action. In 1989, Robin Holliday proposed that the antiaging action of DR evolved in response to the unpredictable periods of food shortage in the wild; he postulated that during such periods of food shortage, animals with genomes that directed resources away from reproduction towards somatic maintenance would have an increased ability to survive the food shortage and thus have a selective advantage. He further hypothesized that when undergoing sustained food shortage (such as the DR regimen in the laboratory) such animals would continue to direct resources towards maintenance and, as a result, the rate of aging would be slowed. This concept has been developed further by the finding that DR protects rats and mice from the damaging effects of acute, intense stressors such as surgery, inflammatory agents, heat, and toxic chemicals. Thus it is reasonable to assume that it also protects against the long-term, low-intensity stressors of the "public" mechanisms of aging. Indeed, progress has been made in uncovering some of the possible mechanisms underlying the protection against stressors. It has been found that DR increases the expression of heat shock proteins in response to stressors, and that it also leads to elevated daily peak plasma concentrations of glucocorticoids; both of these actions may well increase the ability of the organism to cope with stressors.

Thus it is clear that DR has a powerful antiaging action in laboratory rodents, and several mechanisms that may be responsible for its antiaging action have been uncovered. The relative importance of these potential mechanisms remains to be determined.

The question often asked is whether DR also slows the aging processes in humans. Unfortunately, this question cannot be answered, since an appropriate study has never been done, nor is one likely to be done. Indeed, because of the high cost of such studies on large, long-lived species, studies on mammalian species have been completed only on rats, mice and hamsters. Meanwhile, some studies on nonhuman primates have been ongoing for 5 to 10 years, though the results will not be conclusive for another 10 to 15 years. Theoretical reasons have been cited for and against DR's possible antiaging action in humans, but to my mind, the issue is moot. From a practical point of view, few people would be likely to adhere to such intensive dietary restriction over an extended period involving much of the life span. Moreover, such a marked reduction in food intake poses the threat of malnutrition because of possible deficiencies in essential nutrients, which is not a problem in laboratory rodents fed a scientifically designed diet. Thus individuals undergoing this intervention would require careful guidance from and monitoring by a health care professional trained in human nutrition. However, uncovering the mechanisms underlying the antiaging action of DR in rodents may well lead to interventions that would be useful for people.

PROPOSED PHARMACOLOGICAL INTERVENTIONS

The ideal intervention would be a substance taken by mouth that retards aging processes in the broad fashion just depicted for DR. Over the years, claims have been made that many substances have such an action; most of these are not mentioned today because, with time, even the most ardent proponents were disaffected by the absence of evidence of efficacy. However, there are several pharmacological interventions that are currently popular with at least some members of the population; it is worth considering what has led to their use and the evidence of their efficacy. (A pharmacological intervention refers to the administration of substances as drugs, including chemicals not normally present in the body as well as substances normally present but at much higher than usual levels, e.g., a hormone or dietary substance.)

Antioxidants

The possible role of oxidative stress as a cause of senescence was considered in Chapters 4 and 5. It was concluded that oxidative stress may well be an important component of the "public" proximate mechanisms of senescence as well as being involved in some "private" mechanisms. Investigators have used rodent models in an attempt to determine if antioxidant supplementation has antiaging actions similar to those observed with DR. Naturally occurring antioxidant compounds, such as vitamin E, and synthetic antioxidants, such as butylated hydroxytoluene (BHT), have been given to rats and mice either as a single agent or in mixtures (e.g., a mixture of vitamin E, vitamin C, BHT, methionine, and selenite). Both these antioxidant compounds and mixtures of them have been found to be much less effective than DR in retarding senescence in these rodent models; they only modestly increase, if at all, the mean length of life of rodent populations, and do not increase the maximum life span of the populations. A carefully executed study that was recently published by John Holloszy is a typical example; in this study, the diet of one group of rats was supplemented with pharmacological levels of vitamin C, vitamin E, BHT, beta-carotene, and a menadione-sodium-bisulfite complex. There was no difference in mean or maximum length of life between the group given the antioxidant supplement and the group that did not receive the supplementation. Thus it seems clear that in rodents, antioxidant supplementation does not have a marked antiaging action.

Nevertheless, antioxidant preparations are being sold for use in pharmacological doses to retard human aging. Indeed, there is epidemiological evidence that indicates that some antioxidants may retard some aspects of human aging. For example, vitamin E in pharmacological doses has been found to lower the incidence of coronary heart disease and to enhance the immune system in elderly people. Supplementation of the diet with selenium has been reported to decrease the incidence of lung, prostatic, and colorectal cancers. Although inadequate dietary intake of vitamin C increases the risk of cataracts, cancer, and coronary heart disease, there is no evidence that megadoses of vitamin C afford further protection from these age-associated diseases. While it has been reported that beta-carotene retards age-associated cardiovascular disease and cancer as well as helping to maintain memory, the validity of these claims is the subject of controversy. It seems safe to conclude only that certain antioxidant compounds may retard specific aging processes, such as that of a particular age-associated disease process, and that

Deprenyl

This drug has been used to treat the age-associated neurological problems of Parkinson's disease, and appears to be helpful. It is known to inhibit monoamine oxidase B, an enzyme that destroys the neurotransmitter dopamine. It has been suggested that deprenyl acts therapeutically by modifying dopamine metabolism in relevant brain regions.

In 1988, a Hungarian scientist reported that deprenyl markedly extended life span when given to old rats. These results were particularly exciting because the extent of extension was much greater than DR initiated in rats at such advanced ages. Since then, Canadian and Japanese scientists have also found that deprenyl extends the length of life of old rats, mice, and hamsters, but the magnitude of this action was much less marked in these studies than in the Hungarian study. Interestingly, deprenyl has also been found to increase longevity when administered to dogs older than 10 years of age. It has been suggested that the antiaging action of deprenyl may not be due to its ability to inhibit monoamine oxidase B, but rather to its ability to function as an antioxidant. In addition, deprenyl has been reported to increase the activity of the enzymes that protect the rat from oxidative damage, but this action is the subject of debate.

In summary, there is provocative evidence that deprenyl has an antiaging action in rats. However, further work is needed before this claim gains the acceptance accorded DR. Although deprenyl appears to be a helpful therapeutic agent in Parkinson's disease, whether it has other antiaging actions in humans has yet to be explored.

Dehydroepiandrosterone (DHEA)

In Chapter 6, it was pointed out that the plasma concentration of the adrenal cortical hormone DHEA progressively decreases with increasing age in both men and women. Based in part on this finding, DHEA has been promoted as the fountain of youth, and it is prominently displayed in health food stores and in many pharmacies.

Administration of DHEA to rodents has been found to retard the age-associated occurrence of diabetes and cancers as well as memory loss. It also reduces autoimmune-related mortality in autoimmune-prone mice. However, DHEA administration also decreases food intake by rodents, and there has been much debate as to whether DHEA directly retards aging in rodents or does so by causing DR. In several studies in which food intake of control (untreated) rodents was reduced to that of those treated with DHEA, no evidence was found for DHEA slowing senescence.

Epidemiological studies have shown an association between low plasma concentrations of DHEA and the risk of breast cancer, obesity, cardiovascular disease, diabetes, osteoporosis, dementia, and all-cause mortality in humans. Administering DHEA to old people to produce plasma levels similar to those found in the young has been claimed to improve immune function, increase plasma insulin-like growth factor levels, and improve physical and psychological well-being.

Such claims aside, there are also potentially negative aspects of DHEA treatment. High doses of DHEA can cause liver damage. Also, DHEA can be converted in the body to both estrogen and testosterone, and the extent of these conversions is unpredictable; excess estrogen is a risk factor for breast cancer, and excess testosterone may promote prostatic cancer in men and produce a masculinizing effect in women.

Melatonin

The pharmacological use of the hormone melatonin as an antiaging substance has become popular in recent years. As discussed in Chapter 6, it has been shown that melatonin functions as an antioxidant, and that the plasma concentration of melatonin falls with increasing age in both men and women. These findings, coupled with the results of studies using rats and mice, are the basis for this putative antiaging intervention.

It has been reported that adding melatonin to the drinking water of rats and mice increases their length of life. However, these findings must be viewed with caution, because the number of animals studied was small, and data on food intake were not reported. It is of utmost importance that these intriguing studies be repeated with a larger number of animals, and that food intake be carefully monitored to rule out DR as the factor responsible for the results.

It is also important to realize that the efficacy of melatonin as an antiaging agent in humans has not been well studied. Melatonin is sold without prescription, and thus the amount a person decides to take could result in plasma levels many times those normally found in young people. Indeed, not only has the efficacy of melatonin as an antiaging pharmacological intervention in humans not been established, but the possible long-term adverse effects of high plasma levels have yet to be properly evaluated.

Growth Hormone

There is a progressive decrease with age in growth hormone and the plasma level of insulin-like growth factor I, the secretion of which is controlled by growth hormone. These facts led Daniel Rudman and his colleagues to study the effects of treating elderly people with human growth hormone. They studied 21 healthy men who were over 60 years of age and had low plasma levels of insulin-like growth factor I. The subjects were injected with human growth hormone for 6 months, and they had a 9% increase in lean body mass, a 14% decrease in fat mass, a 1.6% increase in lumbar vertebral bone density, and a 7% increase in skin thickness. However, some of these positive effects tended to disappear when the study was extended for another 6 months. A more recent study with healthy elderly men showed that growth hormone treatment modestly increased muscle mass and decreased fat mass, but did not improve muscle strength, endurance, or mental ability. In another study involving women ranging in age from 66 to 82, growth hormone treatment increased muscle mass. The foregoing preliminary results require further study to clearly establish the benefits of this treatment. In addition, there are concerns about potential negative effects, such as the development of diabetes, hypertension, and carpal tunnel syndrome. Moreover, growth hormone must be prescribed by a physician, given by injection (it is a protein destroyed by digestive processes if taken orally), and is expensive (about $15,000 per year). Thus, much research remains to be done before the use of growth hormone as an aging intervention can be recommended.

Estrogen

In Chapter 6, it was pointed out that estrogen replacement therapy often attenuates undesirable symptoms that appear at the time of menopause. In addition, there have been claims of several long-term benefits from this treatment.

There is strong evidence that estrogen replacement therapy retards postmenopausal bone loss. In this way, it reduces the rate of bone fractures due to osteoporosis.

Several epidemiological studies have indicated that estrogen replacement therapy blunts the age-associated progressive increase in coronary heart disease during the postmenopausal period of life. However, a recent clinical trial does not provide support for those studies. In the clinical trial, all of the subjects were postmenopausal women who had a history of coronary heart disease. One group was given estrogen plus a synthetic analogue of progesterone for a 4-to-5-year period. The women in this group had a similar rate of coronary heart disease episodes, such as angina and infarcts, as the women in the control group who had not been so treated. Several questions arise in regard to the apparent discrepancy between this clinical trial and the epidemiological studies. In the epidemiological studies, the women were not assigned to the estrogen therapy group, but rather voluntarily chose this therapy. Were the women who chose the therapy less likely to have coronary heart disease even if they had not undergone the therapy? (The potential for such bias in subject "selection" is an inherent pitfall in this kind of epidemiological study.) Was the absence of protective effect noted in the clinical trial due to the use of a progesterone analogue in addition to estrogen? Would the clinical trial have found estrogen treatment to be efficacious if the subjects of the study had been women with no history of coronary heart disease? Obviously, answers to these and a number of other questions require further research.

There is also suggestive evidence that estrogen therapy may improve memory, both in healthy elderly women and in those suffering from mild Alzheimer's disease. It is of utmost importance that a carefully controlled study on a large number of subjects be done to confirm or deny these preliminary findings on mental function. In addition, the effect of estrogen treatment on cognitive function in men should be investigated.

The potential harmful effects of estrogen therapy must be noted. It increases the risk of uterine endometrial cancer. This risk can be eliminated if, in addition to estrogen, progesterone or a synthetic analogue of progesterone is given for 10 or more days per month. However, this combined hormone therapy results in monthly 3-to-

4-day periods of vaginal discharge of blood and endometrial debris. Estrogen replacement therapy also increases the risk of breast cancer, and it is not clear that the combined estrogen-progestin therapy reduces that risk. In addition, estrogen therapy increases the risk of blood clots in the veins and gallbladder disease in women, while it produces an unwanted feminizing effect in men. Although pharmaceutical companies have been searching for drugs with the beneficial effects of estrogen without its adverse actions, as of now, no such drug has been found.

The decision to use estrogen replacement therapy must be based on the characteristics of the individual. If a woman has a strong family history of breast cancer, then it is probably not wise to choose this therapy. However, if future studies clearly establish that estrogen decreases the risk of coronary heart disease and dementia, then there will be strong reasons to use the therapy. Obviously any decision should be based on the evidence that, in a particular case, the potential for beneficial action outweighs the likelihood of adverse effects.

Testosterone

The plasma concentration of testosterone decreases with age in most men, which has led to the suggestion that testosterone replacement therapy might have an antiaging action in men. A 3-month study of a small group of elderly men with low plasma testosterone levels showed that administration of testosterone results in some increase in lean body mass. These findings indicate that further carefully designed studies on a large number of subjects are warranted. Although it might be expected that administration of testosterone would increase bone and muscle mass, there is also the possibility of negative effects on the prostate, such as promoting the growth of those prostatic cancers that are so slow-growing they would not cause a problem in the absence of this intervention. Thus, much remains to be learned before testosterone replacement therapy can be recommended.

LIFESTYLE AND ENVIRONMENTAL FACTORS

Chapter 6 focused on the many deteriorative changes that characterize the human aging phenotype. However, it was pointed out that there is much individual variation in the age of occurrence of the deterioration if, it must be added, it occurs at all. Rowe and Kahn, in their recent book *Successful Aging*, discuss this heterogeneity and define successful aging in terms of the ability of the individual to maintain the following characteristics into advanced ages: 1. low risk of disease and disease-related disability; 2. high mental and physical functioning; and 3. active engagement with life. They acknowledge that successful aging is, in part, determined by an individual's genes, but they provide substantial evidence in support of their belief that lifestyle and environment play a major role in attaining successful aging. Clearly, lifelong optimal environment and lifestyle are the most effective means of achieving successful aging. However, modification of lifestyle and environment, even at advanced ages, can be remarkably beneficial.

Exercise

Physical fitness refers to the maximum level of exercise that an individual is capable of carrying out. It is often measured by physiologists as the ability of an individual to increase his or her rate of oxygen consumption during a treadmill exercise of increasing intensity. Physical fitness decreases to varying degrees with increasing age. It does not decrease markedly in those who are healthy and have engaged in vigorous physical activity for most of their life. However, many people adopt an increasingly sedentary lifestyle with increasing age; and, in those people, a marked age-associated decrease in physical fitness occurs. Of course, some age-associated diseases also decrease the capacity to exercise, and the subsequent decreased physical activity further decreases physical fitness.

The physically fit have less risk of mortality during the next year than someone of the same age and gender who is sedentary. Moderate exercise, such as walking, is nearly as effective in reducing the risk of mortality as vigorous exercise, such as running the marathon, provided that the moderate exercise program is a sustained one (i.e., daily or several times a week year after year). It has also been shown that old people who have been sedentary for some time can become more physically fit by engaging in an exercise program, and such programs also increase the ability of older people to carry out daily activities.

Both aerobic exercise and resistance training benefit the elderly. Aerobic exercise refers to activities, such as rapid walking and jogging, that increase overall endurance and fitness by promoting cardiovascular and respiratory function. Resistance training, such as weight lifting, increases the size and strength of skeletal muscles, and also provides important daily living benefits for the elderly. These include: improved balance, thereby reducing the risk of falls and bone fractures; increased speed and steadiness when walking; and increased ability to climb stairs. In addition, exercise (aerobic and resistance) reduces the likelihood of developing hypertension and decreases the blood pressure of those suffering from hypertension; reduces the extent of abdominal obesity (potbelly); and increases blood levels of HDL. These actions are probably the main reason that exercise reduces the risk of coronary heart disease and stroke. Exercise also decreases the occurrence of Syndrome X and Type II diabetes.

Compared to DR, exercise in mice and rats has not been found to have dramatic effects on either life span or the breadth of antiaging actions. Nevertheless, the effectiveness of exercise in retarding the deterioration associated with the human aging phenotype is so marked that it is surprising it has not been embraced by more of the people seeking the "fountain of youth."

Of course, exercise is not without risks, but these can be minimized. For example, jogging can cause joint damage; however, walking, which has benefits similar to jogging, rarely causes joint damage. Those with disease problems should receive guidance from a health care provider before adopting an exercise program. For example, some forms of exercise can exacerbate osteoarthritis, but a properly designed program is not damaging, but rather is beneficial. Also, those who are at risk for coronary heart disease or stroke require guidance from a physician to enable them to reap the benefits of exercise with minimum risk.

Diet

In rodents, reducing intake of calories markedly slows the aging processes. Does it have the same effect in humans? Although there is no evidence that decreasing caloric intake in humans influences aging by the same mechanisms as in rodents, it is clear that by decreasing the intake of calories, overweight people experience beneficial effects. These include lowering the blood levels of glucose, lipids, and insulin as well as reducing blood pressure. In addition, caloric reduction increases plasma HDL levels. By these means, the risk of coronary heart disease and stroke are decreased. Interestingly, it appears that most of the effects noted with reduced caloric intake in humans may be secondary to the decrease in abdominal fat content.

Decreasing dietary fat intake is also beneficial. First of all, it facilitates efforts to decrease the intake of calories, because a gram of fat contains more than twice as many calories as a gram of carbohydrate. Also, animal fats, in particular, cause plasma LDL (a risk factor for coronary heart disease and stroke) to increase. In addition, there is evidence that fat, particularly that found in meat and dairy products, promotes the occurrence of colorectal cancers, and there is a hint that this may be true, too, for cancer of the prostate.

Those complex carbohydrates that contain soluble fiber help ward off aging problems; they reduce the formation of colonic diverticuli and decrease blood glucose and blood lipid levels. Fruit, peas, beans, and lentils contain goodly amounts of these carbohydrates.

An increase in dietary protein may benefit many of the elderly, particularly those suffering from an infection or recovering from surgery. About 12% of the caloric intake should come from protein, which can be most easily accomplished by a diet that includes substantial amounts of one or more of the following: meat, fish, poultry, eggs, and dairy products. Of course, high protein intake is not advisable for those with kidney disease.

The dietary intake of calcium by the elderly tends to be less than what is necessary to minimize age-associated bone loss. Many nutritionists believe that about 1200 mg of calcium per day are needed, though most elderly consume only about 700 to 800 mg. Moreover, the elderly show decreased intestinal absorption of calcium. Vitamin D promotes the intestinal absorption of calcium; although diet is one source of vitamin D, much of it is generated by the action of sunlight on the skin. Since many elderly have only limited exposure to the sun, it would be helpful for them to adopt a diet that provides ample amounts of both calcium and vitamin D. Milk fortified with vitamin D is an excellent source of both, provided, of course, that the individual has no problem with lactose intolerance. If the needed amounts of calcium and vitamin D cannot be easily met by diet, a dietary supplement containing these substances in pill or other form is a ready solution.

In those elderly suffering from atrophic gastritis, there is a decreased ability of the small intestine to absorb vitamin B_{12}. This predisposes them to vitamin B_{12} deficiency which, if severe, can lead to pernicious anemia and neurological disorders. For such individuals, dietary sources of vitamin B_{12}, such as meat, will not suffice; they will require a daily dietary supplement of about 1 mg of vitamin B_{12}.

It is particularly important for the elderly to have an adequate intake of folic acid because a deficiency causes high plasma levels of homocysteine, a risk factor for coronary heart disease and stroke. Leafy vegetables, fruit, liver, and yeast are good sources of folic acid, but a dietary supplement to ensure an adequate intake may be a wise choice.

Smoking

While the effect of smoking on the aging of the skin has already been discussed in Chapter 6, smoking also promotes many other aspects of the aging phenotype. Its role as a risk factor for lung cancer was evident more than 30 years ago, and since then it has been found to be a risk factor for other age-associated lung diseases as well as other cancers, such as colorectal cancers and cervical cancer. Probably the most serious effect of smoking on the aging individual is as a risk factor for coronary heart disease; a person who smokes a half pack of cigarettes a day is twice as likely to suffer from coronary heart disease as a nonsmoker of the same age and gender, and smoking one pack of cigarettes a day increases the risk to four times greater. Smoking also increases the risk of stroke. In addition, the use of cigarettes increases the rate of age-associated bone loss, and thus predisposes the smoker to osteoporosis.

Of course, those who have never smoked are the best off in regard to these particular age-associated problems. However, quitting smoking even at advanced ages reduces the risk of most of them. The risk of coronary heart disease and stroke decreases almost as soon as smoking is stopped, and in 5 years an ex-smoker is not much more likely to suffer from these diseases than a person who has never smoked. The risk of lung diseases also decreases after the cessation of smoking, but much more slowly than is the case for cardiovascular diseases. After a smoker has stopped, it takes about 15 years before the risk of lung cancer is similar to that of lifetime nonsmokers. However, it is of interest that the risks due to smoking may be overcome, in part, by exercise regimens that maintain a high level of physical fitness.

Education and Social Support

The likelihood that cognitive function will be maintained into advanced ages increases with the amount of education the individual had in his/her youth. Clearly, education has a lasting impact. Moreover, there is reason to believe that education at advanced ages also enhances cognitive function. For example, elderly people who have memory problems or difficulties in reasoning can greatly improve their performance by participating in training programs for that particular problem, and the improvement is sustained over time.

Social support appears to play a very important part in the quality of physical and mental functioning of the elderly. It has been found that those who do not have close family and/or close friends are more likely to become ill and to have a relatively short length of life. The best predictors of the well-being of an elderly person are the frequency of visits with family and friends and the extent of participation in organizations.

The nature of the social support is critical. Assistance must be given judiciously, or it can be counterproductive. For example, assisting a person with a task that he or she is able to do may decrease the individual's confidence in his/her ability to do the task. Whenever possible, the assistance should be rendered by encouraging the person to do the task for himself/herself.

Although the benefits of social support and education in maintaining functional ability in the elderly are clearly established, the underlying physiological mechanisms for the beneficial actions are not known. Uncovering the biological basis of these effects would not only be intellectually satisfying, but might well yield one of the most powerful interventions for the retardation of aging.

ADDITIONAL READING

Bartke, A., Brown-Borg, H. M., Carlson, J., Hunter, W. S., & Bronson, R. T. (1998). Does growth hormone prevent or accelerate aging? *Experimental Gerontology, 33,* 675–687.

Bellino, F. L., Daynes, R. A., Hornsby, P. J., Lavrin, D. H., & Nestler, J. E. (Eds.) (1995). Dehydroepiandrosterone (DHEA) and aging. *Annals of the New York Academy of Sciences, 774,* 1–350.

Bernaducci, M. P., & Owens, N. J. (1996). Is there a fountain of youth? A review of current life extension strategies. *Pharmacotherapy, 16,* 183-200.

Grodstein, F., Stampfler, M. J., Colditz, G. A., Willett, W. C., Manson, J. E., Joffe, M., Rosner, B., Fuchs, C., Hankinson, S. E., Hunter, P. J., Hennekens, C. H., & Speizer, F. E. (1987). Postmenopausal therapy and mortality. *New England Journal of Medicine, 336,* 1769–1775.

Kitani, K., Kanai, S., Ivy, G. O., & Carrillo, M. C. (1998). Assessing the effects of deprenyl on longevity and antioxidant defenses in different animal models. *Annals of the New York Academy of Sciences, 854,* 291–306.

Masoro, E. J. (1998). Hormesis and the antiaging action of dietary restriction. *Experimental Gerontology, 33,* 61–66.

Reiter, R. J. (1995). The pineal gland and melatonin in relation to aging: A summary of the theories and of the data. *Experimental Gerontology, 30,* 199–212.

Rowe, J. W., & Kahn, R. L. (1998). *Successful aging.* New York: Pantheon.

Weindruch, R., & Sohal, R. S. (1997). Caloric intake and aging. *New England Journal of Medicine, 337,* 986–994.

Yu, B. P. (1995). Putative interventions. In E. J. Masoro (Ed.), *Handbook of physiology, Section 11, Aging* (pp. 613–631). New York: Oxford University Press.

Index

Acid-base balance, 143–144
  acids, 143
  bases, 143
  carbonic acid concentration, 143
  chemical buffers, 143
  metabolic acidosis, 144
  renal regulation, 143–144
  respiratory regulation, 143
Actin, 94
Activities of Daily Living (ADL), 29, 100, 107
Adipocytes, 87, 90
Adipose tissue, 86–87, 89–90, 155
  mass, 156
Adrenal cortex, 70, 141–142, 158–159
Adrenal glands, 68
Adrenal medulla, 117, 158
Adrenocorticotrophic hormone (ACTH), 158–159
Aerobic organisms, 53–54
Age-associated disease, 4–5, 82–83, 85, 137
Age-specific death rate, 2–3, 10, 15, 18–22
  gender differences, 25–26
Age-structure of population, 13–15, 22–26
  gender composition, 25–26
  influence of birth rate, 22–25
  influence of life expectancy, 22–24
  influence of migration, 22–23

Aging
  definition of, 1
  environmental factors, 82–83
  genetic clocks, 59
  phenotype, 40–41, 47–48, 62, 72, 75, 85–169
  "private" mechanisms, 77–80, 82, 84
  proximate biological mechanisms, 33, 47–72, 75, 77–83
  "public" mechanisms, 77–80, 82, 84
  unifying concept of biological basis, 79–84
Aldosterone, 141–142, 153
Alzheimer's disease, 5, 9, 56, 78–79, 83, 107, 121–123, 180; *see also* Dementia
  alpha$_2$-macroglobulin allele, 123
  apolipoprotein-E, epsilon-4 allele, 112–123
  beta-amyloid, 122
  estrogen replacement therapy, 123
  familial, 122
  folic acid, 123
  neurofibrillary tangles, 122
  nonsteroidal anti-inflammatory drugs, 123
  psychosocial factors, 123
  senile plaques, 122
Amino acids, 60, 65

Amyotrophic lateral sclerosis, 5, 148
Anatomical changes, 3–4, 85
Anemia, 137
   pernicious, 151, 185
Angiotensin II, 141
Antiaging interventions, 171–186
   dietary, 171–175
   lifestyle and environmental factors, 181–186
   pharmacologic, 175–181
Antioxidant compounds, 54, 81, 159, 175–178
Apolipoproteins, 156
   apolipoprotein E, ε4-allele, 78
Apoptosis, 61–63, 166
Arteries, 129–132; *see also* Blood pressure
   impedance, 130
   lumen diameter, 130
   pulse wave, 130
   wall stiffness, 130
   wall thickness, 130
Atherosclerosis, 9, 56, 82, 129–130, 156, 174
ATP (adenosine triphosphate), 53, 57, 62, 80, 95, 106, 154
Austad, S. N., 43–44, 77
Autoimmune diseases, 71–72, 100, 132–133, 168, 172
Autonomic nervous system, 117–118

"Baby Boom Generation," 25, 29–30
Beta-carotene, 176
Biological age, 5–6, 30
Biomarkers of aging, 5–6, 22
Blood clots, 129–130, 181
   embolus, 129
   thrombus, 129
Blood pressure, 85, 129–133, 184; *see also* Hypertension
   baroreceptor reflex, 132
   diastolic pressure, 131
   mean pressure, 131
   postprandial hypotension, 132–133
   postural hypotension, 132–133
   systolic pressure, 131
Blood sugar, 85, 156
Blood volume, 142
Body composition, 86–89, 154
Body fat, 156–158
   content, 85, 87
   distribution, 85, 87, 157, 164
   visceral, 87, 184
Body structure, 86–89
Body weight, 86
Bone, 94, 96–98
   density, 97–98, 179
   formation, 98
   fractures, 97–98, 180, 183
   loss, 1, 70, 78, 97–98, 143, 180, 185
   osteoblasts, 96
   osteoclasts, 96–97
   osteocytes, 96–97
   remodeling, 97
Bone marrow, 63, 165, 167
Brain, 62, 88, 95, 101–102; *see also* Nervous system
   age-associated cognitive disorders, 121–124
   attention, 119
   blood flow, 106
   brain stem, 96, 101, 117, 135, 144
   cerebellar degeneration, 115
   cerebral cortex, 101–102, 107, 118, 122, 144
   cerebrovascular disease, 121
   cognitive functions, 73, 110, 119–124, 185
   corpus striatum, 101–102, 116–117
   creativity, 119, 121
   forebrain, 101, 118, 159
   hippocampus, 70–71, 122

Index

hypothalamus, 69, 101–102, 118, 142–143, 158, 162, 168
intellectual functions, 121
learning, 119, 121
medulla, 101–102
memory, 117, 119–122, 176, 178, 180, 185
metabolism, 106
midbrain, 101
perception, 119
pons, 101–102
semantic knowledge, 121
sensations, 107
substantia nigra, 116
thalamus, 101, 119
thinking, 119
volume, 105
weight, 105
Brown-Sequard, C. E., 68

Calcium, 94, 96, 98
Caloric restriction (CR), *see* Dietary restriction (DR)
Cancer, 4–5, 9, 30, 62, 64, 71, 88, 159, 165, 168, 172
  breast, 162, 178, 181
  colon, 62
  colorectal, 176, 184–185
  leukemias, 167
  lung, 176, 185
  lymphoma, 167
  prostate, 5, 78, 164, 176, 178, 181, 184
  skin, 91, 93
  uterine cervix, 162, 185
  uterine endometrium, 162, 180
Carbohydrate metabolism, 155–157
  fructose metabolism, 155
  galactose metabolism, 155
  glucose metabolism, 129, 155–157, 183–184
  glucose tolerance test, 155
  glycogen, 155
  impaired glucose tolerance, 156
  insulin resistance, 156–157
Carbon dioxide, 53
Cardiovascular system, 68, 117, 121–134, 153, 158, 168; *see also* Arteries; Heart; Microvasculature; Veins
  disease, 64, 87, 128, 159, 176
  pulmonary circuit, 124–125
  systemic circuit, 124, 133–134
Cartilage, 98–100
Casals, P., 121
Cell culture model of aging, 63–64
Cells, 6, 50, 58–59, 61, 67–68, 86–89, 134
  atrophy, 88
  cytoplasm, 50
  digestion vesicles, 51
  endoplasmic reticulum, 51
  Golgi apparatus, 51
  hyperplasia, 88
  hypertrophy, 88
  loss, 61, 63, 88
  migration, 58
  membranes, 50, 53–54
  nuclei, 7, 50, 60
  proliferation, 58, 64, 93
  transport vesicles, 51
Cellular signal transduction, 69
Central nervous system, *see* Nervous system
Cerami, A., 55
Charlesworth, B., 39
Cholesterol, 58
Cholesterol metabolism, 156
Chromosomes, 7–9, 35, 61–62, 123
Cohort, definition of, 3
Collagen, 58, 91–92, 96, 130
  cross-linking, 56
Colon, 149–150
  gastrocolic reflex, 149
  salt absorption, 153
  water absorption, 149, 153

Connective tissue, 89
Coronary heart disease, 5, 78–79,
    82–83, 128–129, 156–157,
    176, 180–181, 183–185
  ischemia, 129
  myocardial infarction, 129
Crick, B. F., 34
Cutler, R., 65
Cytokines, 107

Darwin, C., 34, 37
Defecation, 146, 150
  anus, 146, 149
  constipation, 150
  diarrhea, 150, 153
  external anal sphincter, 150
  fecal incontinence, 93, 150
  feces, 149
  internal anal sphincter, 149
  rectum, 150
De Leon, P., 171
Dementia, 107, 117, 121–124, 159,
    178; *see also* Alzheimer's
    disease
  mixed Alzheimer-cerebrovascular
    disease, 124
  multi-infarct, 123
  transient ischemic attack, 123
  white matter lesions, 123
Deprenyl, 177
  increased rodent longevity,
    177
  therapy for Parkinson's disease,
    177
DHEA (dehydroepiandrosterone),
    69, 159
  decreased rodent food
    consumption, 178
  increased rodent longevity, 178
  immune function, 178
  plasma concentration, 159
  source of estrogen, 159, 178
  source of testosterone, 159, 178

Diabetes mellitus, 5, 55–56, 123,
    129, 145, 159, 178–179
  Type II, 157, 183
Diet, 9, 131, 183–185; *see also*
    individual vitamins
  calcium, 184
  caloric intake, 183–184
  complex carbohydrates, 184
  eggs, 184
  fish, 184
  fruit, 143, 184–185
  meat, 184
  milk, 184
  poultry, 184
  vegetables, 143, 184–185
Dietary fuels
  carbohydrates, 53, 55
  fats, 53, 184
  proteins, 53, 184
Dietary restriction (DR), 171–175,
    178, 183
  age-associated disease slowed,
    172
  age-specific death rate
    decreased, 172
  body fat decreased, 173
  caloric restriction, 173
  heat shock protein expression
    increased, 174
  length of life increased, 172
  metabolic rate unchanged, 173
  mortality rate doubling time
    (MRDT) increased, 172
  oxidative damage decreased, 173
  physiological deterioration
    slowed, 172
  plasma glucocorticoid
    concentration increased,
    174
  plasma glucose concentration
    decreased, 173
  plasma insulin concentration
    decreased, 173

protection against stressors, 174
reactive oxygen molecule
  generation decreased, 174
Differentiation, 65
Disposable Soma Theory, 77, 79, 81
DNA, 7, 9, 34–35, 52–53, 55–56,
  60–62, 64–67
  chromosomal, 63, 80
  damage, 61–63, 81
  methylation, 66
  mitochondrial, 62, 80
  repair, 54–55, 62, 81
Down Syndrome, 8–9, 122, 160
Dysdifferentiation Theory, 65

Ear, 107, 108–110; *see also* Hearing
  basilar membrane, 109
  cochlea, 109
  eardrum (typanum), 109
  hair cells, 109
  incus, 109
  malleus, 109,
  stapes, 109
  outer ear, 109
  oval window, 109
  vestibular portion, 109
  wax 109–110
Edema, 134
Elastin, 58, 91–92, 130
Electroencephalogram (EEG), 118
Endocrine system, 57, 68–69,
  153–160
Energy metabolism, 154–155
Entropy, 2
Enzymes, 53
Epinephrine (adrenaline), 117
Error Catastrophe Theory, 65
Esophagus, 148
  lower esophageal sphincter, 148
  upper esophageal sphincter, 148
Estrogen replacement therapy,
  179–181
  Alzheimer's disease, 123, 180

blood clots
bone fracture, 180
bone loss, 180
breast cancer, 181–182
coronary heart disease, 180–181
gallbladder disease, 181
memory, 180
uterine endometrial cancer, 180
Evolutionarily adaptive aging
  clocks, 48–49, 59, 76
Evolutionary biology, 33–37;
  see also Natural selection
  evolutionary forces, 2, 33
  fitness, 35–36, 77
Evolutionary theories of aging,
  history, 37–38
Evolutionary theory of aging, 33,
  38–44, 48–49, 59, 75–76,
  79–80, 83
  genetic mechanisms, 41–42, 59,
    75
  supporting evidence, 42–44
Exercise, 4, 98, 100, 131–132, 136,
  155, 182, 186
  aerobic, 183
  resistance, 95, 183
Extracellular fluid, 57
Extracellular matrix, 58–59, 88–89,
  91–93, 96
Eye, 107, 111–113; *see also* Vision
  age-related macular degeneration,
    113
  aqueous humor, 111
  cataracts, 56, 112, 176
  ciliary muscles, 111
  color vision, 112
  cones, 111
  cornea, 111
  glaucoma, 112
  lens, 55–56, 111
  power of accommodation, 111
  presbyopia, 111
  pupil, 111

Eye *(continued)*
  retina, 111–113
  rods, 111–112
  vitreous body, 111

Falls, 114, 183
Fat metabolism, 156–157
  plasma lipids, 184
Fatty acids, 57
  peroxidation of polyunsaturated, 54
  unsaturated, 58
Fecundity, 10, 17, 40
Fibroblasts, 63
Fischer, K. E., 43–44
Fischer, R. A., 38
Folic acid, 185
  Alzheimer's disease, 123
  homocysteine, 185
Frailty, 83
Free Radical Theory, 52–53
Free radicals, 52–54
Fuel utilization, 52–53, 55, 95
  carbohydrate, 55, 57, 77, 80, 154, 158
  fat, 77, 154, 158
  glucose, 106
  protein, 77, 154

Gallbladder, 151–153
  cholecystokinin (CCK), 151–153
  contraction, 151
  disease, 181
  gallstones, 151–152
Gametes, 10, 36, 59
Gastrointestinal System, 132, 146–153
  absorption, 152–153
  digestion, 152
  motor activity, 146–150
  secretion, 150–152
Gene-environment interactions, 6–7, 9

Gene expression, 48, 59–62, 64–67
  transcription, 60, 65–67
  translation, 60, 65–67
Genes, 6–9, 34–36, 39–40, 47–48, 54, 56, 59–62, 65–66, 79, 182
  alleles, 7–8, 36
  heat shock protein 70 gene, 67
  helicase gene family, 9
  late-life deteriorative actions, 82
  mutations, 9, 35, 62, 71, 78
  stress response genes, 158
Genetic drift, 36, 78
Genetic mechanisms of aging
  antagonistic pleiotropy, 41–42, 49, 59, 76, 78–79
  mutation accumulation, 41, 59, 76, 78
Genetics, 6–9, 42–43, 49, 60, 136, 152
  abnormalities, 160
  heritability, 8, 42
  heterozygosity, 8
  homozygosity, 8
  Mendelian, 34
  molecular, 34–35
  population, 38
  transgenic animals, 54
Genome, 6, 76
Germ line, 10, 64
Glucocorticoid Cascade Hypothesis, 70–71
Glucocorticoids, 70–71, 158–159
  cortisol, 68, 158–159
Glycation, 55–56, 59, 80
  advanced glycation end-products (AGE), 55
  Amadori product, 55
  Schiff base, 55
Glycation Hypothesis, 55–57
Glycoxidation, 80
Glycoxidation Hypothesis, 56–57
Golgi tendon organ, 108
Gompertz, B. 19

Index

Gout, 5, 101
Growth hormone, 68–69, 157, 179
  bone density, 179
  diabetes mellitus, 179
  hypertension, 179
  lean body mass, 179
  muscle mass, 179
  muscle strength, 179
  plasma growth hormone, 179
  plasma insulin-like growth factor I (IGF-I), 179
  skin thickness, 179

Hair
  distribution, 85
  graying, 85–86
  hair follicles, 90–91
Haldane, J. B. S., 38
Hamilton, W. D., 39
Harman, D., 52
Hayflick, L., 63–64
"Hayflick limit," 63–64
Hearing, 108–110; *see also* Ear
  auditory nerve, 109
  loudness, 108–109
  noise, 110
  pitch, 108–109
  prebycusis, 109–110
  timbre, 109
Heart, 62, 125–130
  action potentials, 126
  arrhythmias, 126
  atrioventricular node, 126
  cardiac output, 126–128, 138
  conductile system, 125–126
  congestive heart failure, 129
  contractility, 128
  coronary circulation, 128–130
  diastole, 126
  electrocardiogram, (EKG) 126
  exercise, 128
  heart rate, 126, 128, 132
  left ventricular end-diastolic blood volume, 128
  left ventricular hypertrophy, 88, 125, 129
  pacemaker cells, 126
  parasympathetic nerve supply, 127–128
  pump function, 126–127
  sinoatrial node, 126
  size, 125
  stroke volume, 126–128
  sympathetic nerve supply, 128
  systole, 126, 131
  valves, 127, 129, 131
Heartburn, 148–149
Heat shock proteins, 67, 81, 158
Height, 86
Hemolysis, 71
Heydari, A., 67
Histamine, 93
Holliday, R., 174
Hormone receptors, 69
Hormones, 68–69, 107, 153
Huntington's disease, 78, 116–117
Hutchinson-Gilford Syndrome, 8
Hydrogen ion concentration, plasma, 143–144
Hydrogen peroxide, 53
Hydroxyl radical, 53
Hypertension, 123, 125, 129, 131–132, 140, 157, 179, 183
  Isolated Systolic Hypertension, 131
Hypothalamic-pituitary system, 70

Immune system, 9, 63–64, 68, 70–71, 158, 165–168
  adenoids, 165
  antibodies, 166–167
  antigens, 166–167
  B-lymphocytes, 165–167
  cellular immunity, 166
  humoral immunity, 166

Immune system *(continued)*
  infectious diseases, 167
  lymph nodes, 165
  lymphoid tissues, 167
  lymphokines, 166
  macrophages, 165–166
  neutrophiles, 165–166
  Peyer's patches, 165
  phagocytosis, 166
  plasma cells, 166
  thymic hormones, 167
  thymus, 165–167
  T-lymphocytes, 165–167
  tonsils, 165
Immune Theory, 71–72
Infectious agents, 80
Inflammation, 100–101, 158
Instrumental Activities of Daily Living (IADL), 28–29
Insulin, 68, 155–157
Insulin-like growth factor I (IGF-I), 157, 178–179

Johnson, T., 77
Joints, 94, 98–101, 114–116
  capsule, 98, 108
  damage, 183
  flexibility, 99–100
  ligaments, 98, 108
  synovial fluid, 99
  synovial membrane, 99

Kahn, R. L., 4
Keratinocytes, 89–91
Kidneys, 137–144, 153
  acid-base balance, 143–144
  basement membrane thickening, 56
  body fluid composition regulation, 137–138, 141
  body fluid solute concentration regulation, 142–143
  collecting ducts, 137, 140
  concentrated urine, 142
  connecting ducts, 137
  dilute urine, 143
  disease, 172, 184
  extracellular fluid volume regulation, 141–142
  glomular filtration, 140, 142
  glomeruli, 137–139
  glomerulonephritis, 71, 168
  mass, 137
  nephrons, 137–141
  peritubular capillaries, 138, 140–141
  potassium secretion, 142
  renal arteries, 137–138
  renal blood flow, 140, 142
  renal veins, 138
  sodium reabsorption, 142
  tubular reabsorption, 138, 140–141
  tubular secretion, 141
  ureters, 137–138, 144
  water reabsorption, 142–143
Kirkwood, T., 77
Kristal, B., 56

Langerhans cells, 89–92
Lean body mass, 86–87, 179, 181
Length of life
  maximum, 18
  mean, 18
  median, 18
Life expectancy, 15–17, 22, 98
  gender differences, 25
Life span, 54
Life tables, 14–18
  cohort life tables, 17–19
  period life tables, 17–18, 20
Lifestyle, 9, 83, 169, 182
  education, 186
  sedentary, 87, 95, 182
  smoking, 185–186
  social support, 186

Lipids, 50
  peroxidation, 58
Lipoproteins, serum
  high density, 129, 156–157, 183–184
  low density, 82, 129, 156–157, 184
Liver, 146, 151, 155, 157
  bile, 151
  bile pigments, 151
  cholesterol, 152
  micelles, 151
Lungs, 134–136, 142
  alveolar ventilation, 136–137
  alveoli, 134–136
  functional residual capacity (FRC), 134
  residual volume, 135
Lymph, 165

Martin, G. M., 8, 77–79
Mast cells, 93
Mastication, 146, 148
McCay, C., 171
Medawar, P., 38–39, 41, 61, 76
Medvedev, Z., 65
Meissner corpuscles, 107
Melanocytes, 86, 89–91
Melatonin, 69, 159–160, 178–179
  antioxidant, 178
  circadian rhythm, 159
  pineal gland, 159
  plasma concentration, 159, 178
  rodent length of life, 178
Membrane Hypothesis, 58
Membranes, biological, 57–58
  fluidity, 57–58
  lipids, 80
Mendel, G., 34
Menopause, 97–98, 160–163
  breast changes, 162
  estrogen replacement therapy, 163
  fatigue, 163
  gonadotrophin releasing hormone (GnRH) secretion, 162
  "hot flushes" ("hot flashes"), 163
  irritability, 163
  libido, 163
  oviduct changes, 162
  plasma estradiol, 162
  plasma follicle stimulating hormone (FSH), 162
  plasma luteinizing hormone (LH), 162
  psychological changes, 163
  pubic hair changes, 162
  uterine changes, 162
  vaginal changes, 162
Merkel disks, 107
Metabolic functions, 153–158
Metabolic rate, 51, 154–155
  basal metabolic rate (BMR), 154
  diet-induced thermogenesis, 154–155
  physical activity, 154–155
  shivering thermogenesis, 154
Microvasculature, 89, 132–133
  arterioles, 131–133
  capillaries, 131–133
  exchange vessels, 133
  metarterioles, 132–133
  precapillary sphincters, 133
  resistance vessels, 131–133
  venules, 132
Mitochondria, 51, 53, 57, 62
Mitosis, 63–64, 89
Moorhead, P.
Morgan, T. H., 34
Mortality, 2–3, 14–22, 182
Mortality rate doubling time (MRDT), 21–22
Mouth, 134, 150
Multiple sclerosis, 5
Muscle, cardiac, 94, 117, 129

Muscle, skeletal, 62, 88, 94–95, 99, 101, 108, 113–116, 132, 144, 146, 148, 150, 164
  contractile proteins, 94
  fibers, 94–96
  mass, 87, 95, 155, 157–158, 179, 181, 182
  power, 95
  sarcoplasmic reticulum, 94
  strength, 95, 115, 182
Muscle, smooth, 94, 117, 132–133, 144, 146, 149
  contactile activity, 130
Musculoskeletal system, 94–101
Myasthenia gravis, 148
Myosin, 94

Natural selection, 34–42, 49, 76, 78–79, 83; *see also* Evolutionary biology
  group selection, 36–38
  individual selection, 36–38
Neoplasia, 88
Nerve cells, *see* Neurons
Nerve fibers, 90
Nervous system, 57, 69, 101–124, 133, 158, 168; *see also* Brain
  balance, 114–116, 183
  central processing, 113–114
  cerebrospinal fluid, 105
  glial cells, 104–105, 122
  locomotion, 115–116
  motor disorders, 116–117
  motor functions, 113–117
  myelin, 104
  peripheral nervous system, 101
  posture, 114–115
  reaction time, 114
  sense organs, 107
  sensory functions, 107–115, 118, 120
  spinal cord, 95–95, 107, 117, 144
  synapses, 104–105

Neuroendocrine system, 68–70
Neuromuscular function, 95–96, 113
  motor units, 96
  neuromuscular junctions, 96, 104
Neurons, 65, 69, 101, 103–106
  action potentials, 96
  atrophy, 105–106
  axons, 96, 101, 103–104, 113–114, 122
  dendrites, 101, 103–105, 122
  loss, 105–106
  motor, 96, 114
  nerve impulse, 96, 104
  types, 101, 103
Neurotransmitters, 69, 104, 107
  acetylcholine, 96, 117–118, 128
  dopamine, 116, 177
  norepinephrine, 117–118, 128
Nuland, S. B., 38

Obesity, 87, 159, 178
  abdominal, 157, 183
Organ structure, 58–59, 89
Orgel, L., 65
Osteoarthritis, 100, 115–116, 183
Osteoporosis, 5, 97–98, 159, 178, 185
  risk factors, 98
  Type I, 97
  Type II, 97
Oxidative damage, 54, 81, 159
Oxidative metabolism, 53, 57, 80, 154
Oxidative stress, 53, 56, 58–59, 61, 80, 82–83, 175
Oxidative Stress Hypothesis, 53–57

Pacinian corpuscles, 107
Pain, 107
Pancreas, 68, 146, 151
  amylase, 152
  insulin secretion, 155, 157
  pancreatic juice, 151
Parasympathetic nervous system, 117, 144, 146

Parkinson's disease, 5, 115–117, 148
Pearl, R., 51
Peptic ulcer, 5
Phenotype, 7–8, 34
Physical activity, 156–157, 182
Physical fitness, 156, 169, 182–183, 185
  mortality, 182
Physiological changes, 3–4, 83, 85
Piantanelli, L., 22
Picasso, P., 121
Pituitary gland, 68–69
  anterior lobe, 69, 157, 158, 164
  posterior lobe, 69, 142
Pneumonia, 137, 167
Pollutants, 80
Population aging rate, 18–22
Potassium, plasma, 141
Proprioception, 108–109, 114–115
Prostate, 119, 145, 163–164
  benign prostatic hyperplasia, 88, 146, 165
Protective Processes, 54–55, 80–81
Proteins, 50, 55, 60, 64–66, 80
  altered, 57
  biosynthesis, 66–67, 81, 157
  carbonylation, 54
  degradation, 157
  glycation, 57, 157
  heat damage, 157
  membrane, 57
  metabolism, 157–158
  oxidation, 57, 157
  sulfhydryl groups, 54
  turnover, 57, 66, 157
Proteolytic enzymes, 81

Rate of Living Theory, 51–53
Reactive oxygen molecules, 53–54, 62, 80–81
Red blood cells, 136
Renal and urinary system, 137–146; see also Kidney; Urination

Renin, 141–142
Repair processes, 50, 54, 80–81
Reproduction, 77
Reproductive system, female
  160–163; see also Menopause
  breasts, 160
  corpus luteum, 160
  estrogen, 68–69, 97, 160, 179–181
  fertility, 161
  follicle stimulating hormone (FSH), 160
  luteinizing hormone (LH), 160
  gonadotrophin releasing hormone (GnRH), 160
  menstrual cycle, 160
  menstruation, 160
  ova, 10, 160
  ovaries, 68, 160
  oviducts, 160
  ovulation, 160
  pregnancy, 160
  progesterone, 160, 180
  uterus, 160
  vagina, 160, 165
Reproductive system, male, 163–165; see also Prostate
  fertility, 164
  follicle stimulating hormone (FSH), 164
  hair distribution, 164
  impotence, 165
  libido, 164
  luteinizing hormone (LH), 164
  penis, 163–165
  pitch of voice, 164
  semen, 164
  seminal vesicles, 163–164
  seminiferous tubules, 164
  skeleton, 164
  sperm, 10, 163–164
  testes, 163–164
  testosterone, 69, 78, 163–164, 181

Respiratory system, 134–137, 168; see also Lungs
  breathing frequency, 135
  bronchi, 134
  bronchioles, 134–135
  carbon dioxide transport, 137
  chronic obstructive pulmonary disease, 136–137
  diaphragm, 135
  elastic forces, 134–135
  expiration, 135–136
  FEV_1, 135–136
  gas transport, 136–137
  hemoglobin, 136–137
  inspiration, 135
  lung-thorax pump, 134–136
  oxygen transport, 136–137
  pulmonary function tests, 135–136
  thoracic expiratory muscles, 135
  thoracic inspiratory muscles, 135
  trachea, 134–135
  vital capacity, 135
Rheumatoid arthritis, 71, 100–101, 168
RNA, 60
  messenger RNA, 66
Robinson, M. ("Grandma" Moses), 121
Rose, M. R., 42–44
Rosen, R., 67
Rowe, J. W., 4, 182
Rubner, M., 52, 173
Rudman, D., 179

Salivary glands, 146, 150
  amylase, 150, 152
  parotid, 150
  saliva, 148, 150
  starch digestion, 150, 152
  sublingual, 150
  submaxillary, 150
Sapolsky, R., 70–71

Sebaceous glands, 90–91
Sebum, 91
Segmental progeroid syndromes, 8–9
Selenium, 176
Senescence, definition of, 1
Senile, definition of, 1
Senility, definition of, 1
Sexual maturation, 81
Shakespeare, W., 13
Skin, 63, 89–94, 142, 184
  allergies, 92–93
  barrier function, 89, 92
  basement membrane, 90–91
  delayed hypersensitivity reactions, 92
  dermis, 89–91, 93
  epidermis, 89–91, 93
  extrinsic aging, 91–93
  immediate hypersensitivity reactions, 92
  immune function, 89, 92
  inflammation, 93
  intrinsic aging, 89–91
  photoaging, 92–93
  pressure sores (bedsores), 93–94
  smoking, 91–92
  subcutaneous fat, 89–91
  sun damage, 91, 93
  thickness, 179
  wrinkling, 91–92
Sleep, 118–119, 159, 163
  apnea, 119
  rapid eye movement (REM) sleep, 118–119
  slow wave sleep (SWS), 118–119
Small intestine, 63, 149, 151–153
  absorption of fat digestion products, 153
  absorption of minerals, 153
  absorption of protein digestion products, 153
  absorption of salts, 153

Index

absorption of vitamins, 153
  calcium absorption, 153
  duodenum, 149
  fat digestion, 152
  glucose absorption, 153
  ileum, 149
  jejunum, 149
  lactase, 152
  lactase deficiency, 152
  lactose digestion, 152
  lactose intolerance, 152, 184
  micelles, 152
  protein digestion, 152
  starch digestion, 152
  sucrase, 152
  sucrose digestion, 152
  water absorption, 153
Smell, 113
Societal impact of aging population, 27–29
  daily living assistance, 28–29
  dependency ratios, 27–28
  disease, 28
  work force, 27–28
Sodium, plasma, 141
Soma, 10
Somatic cells, 35, 61, 63–64
Somatic maintenance, 77
Somatic Mutation Theory, 60–62
Speciation, 36
Stomach, 148, 150–152
  atrophic gastritis, 151, 185
  chyme, 149
  gastric emptying, 149
  gastric juice, 149
  hydrochloric acid, 151–152
  intrinsic factor, 151
  parietal cells, 151
  pepsin, 151
  protein digestion, 151
Stress, 70, 118, 158–159
  stress response genes, 158

Stressors, 158
  long-term, low intensity, 80–83
Stroke, 5, 82, 115, 131, 156–157, 183–185
"Successful aging," 4, 182
Sugars, reducing, 55
Superoxide dismutases, 81
Superoxide radical, 53
Survival curves, 18–20
Survival rate, 17
Swallowing, 146, 148–149
  achalasia, 148
  food bolus, 148, 150
Sweat glands, 90–91
Sympathetic nervous system, 117, 133, 144, 146, 158
Syndrome X, 156–157, 183
Systemic function, regulation of, 67–72
Systemic lupus erythematosus, 71, 168
Szilard, L., 60

Taste, 113
Telomerase, 64, 171
Telomere Senescence Theory, 63–64
Telomeres, 63–64
  shortening, 64, 171
Temperature receptors, 107
Tendons, 94, 99, 114–115
Testosterone replacement therapy, 181
  bone mass, 181
  lean body mass, 181
  muscle mass, 181
  prostate cancer, 181
Theories of Aging, 47–72
Thermal insults, 80
Thermodynamics, second law, 2
Thermoregulation, 4, 168–169
  behavioral responses, 168–169
  body heat, 168
  core body temperature, 168

Thermoregulation *(continued)*
  heat production, 168
  heat stroke, 169
  hypothalamic receptors, 168
  hypothermia, 169
  shivering, 168–169
  skin blood flow, 168–169
  skin receptors, 168
  sweat glands, 168–169
Thirst, 143
Thyroid gland, 154
  hypothyroidism, 68, 154
  thyroid hormone, 68, 154
Tissue structure, 58–59, 89
Touch, 107
Transcription, *see* Gene expression
Transcription factors, 66–67
Translation, *see* Gene expression
Tropomyosin, 94
Troponin, 94
Tuberculosis, 167
Tumors, benign, 88

Uric acid, 101
Urination, 144–146, 168
  external urinary sphincter, 144–145
  incontinence, 93, 145–146
  internal urinary sphincter, 144–145
  reflex responses, 144
  urethra, 144–145
  urinary bladder, 137, 144–146
  urine, 137, 142, 144
  urine volume, 142

Vasopressin, 142, 154

Veins, 132–134
  diameter, 134
  distensibility, 133
  valves, 134
Vertebrae, 97–98
Vertigo, 108
Vestibular apparatus, 108, 114–115
  saccule, 108
  semicircular canals, 108
  utricle, 108
Vision 110–115; *see also* Eye
  occipital lobe, 111
  optic nerve, 111, 113
  visual acuity, 111–113
Vitamin A, 54
Vitamin B_{12}, 151, 185
Vitamin C, 54, 176
Vitamin D, 153, 184
Vitamin E, 54, 81, 176

Walford, R., 71–72
Walking, 85, 95, 115–116, 182–183
Wallace, A. F., 34, 37
Water, 53, 142
  intracellular, 88
Watson, J., 34
Wear and tear theories, 48, 50–60, 63
Weismann, A., 37
Werner Syndrome, 8–9
Williams, G. C., 41–42, 76
Wound healing, 64, 93
Wright, F. L., 121

Yu, B. P., 56

Zs-Nagy, I., 58